WOMEN AND CLIMATE CHANGE

WOMEN AND CLIMATE CHANGE

Examining Discourses from the Global North

NICOLE DETRAZ

The MIT Press
Cambridge, Massachusetts
London, England

The MIT Press would like to thank the anonymous peer reviewers who provided comments on drafts of this book. The generous work of academic experts is essential for establishing the authority and quality of our publications. We acknowledge with gratitude the contributions of these otherwise uncredited readers.

This book was set in Adobe Garamond and Berthold Akzidenz Grotesk by Jen Jackowitz. Printed and bound in the United States of America.

Library of Congress Cataloging-in-Publication Data

Names: Detraz, Nicole, author.
Title: Women and climate change : examining discourses from the global North / Nicole Detraz.
Description: Cambridge, Massachusetts : The MIT Press, [2023] | Includes bibliographical references and index.
Identifiers: LCCN 2022022481 (print) | LCCN 2022022482 (ebook) | ISBN 9780262542074 (paperback) | ISBN 9780262362115 (epub) | ISBN 9780262372664 (pdf)
Subjects: LCSH: Women and the environment—Developed countries. | Climatic changes—Social aspects—Developed countries.
Classification: LCC GE195.9 .W478 2023 (print) | LCC GE195.9 (ebook) | DDC 304.2/8082—dc23/eng20221004
LC record available at https://lccn.loc.gov/2022022481
LC ebook record available at https://lccn.loc.gov/2022022482

10 9 8 7 6 5 4 3 2 1

For Ali John—May your generation do what others would not.

Contents

Acknowledgments *ix*

1 **INTRODUCTION: WHAT ARE THE CONNECTIONS BETWEEN GENDER AND CLIMATE CHANGE?** *1*

2 **EMPLOYING A FEMINIST METHODOLOGY** *15*

3 **WOMEN AS VULNERABLE: PRECARIOUSNESS IN THE FACE OF CLIMATE CHANGE** *29*

4 **WOMEN AS CAREGIVERS: CARE ROLES AND CLIMATE CHANGE** *65*

5 **WOMEN AS KNOWLEDGEABLE: EXPERTISE AND COMPETENCE IN CLIMATE CHANGE** *97*

6 **WOMEN AS AGENTS: PARTICIPATION AND BARRIERS IN CLIMATE CHANGE WORK** *133*

7 **CONCLUSIONS: WHERE DO WE GO FROM HERE?** *175*

Appendix: Participant Information *207*
Notes *213*
References *223*
Index *253*

Acknowledgments

I would like to begin by thanking all the women who agreed to speak with me for this book. Thank you for sharing your thoughts with me, and for all the many ways you are working to get us to a greener world. You have given me hope.

Many people contribute to writing academic books—students, colleagues, and editors for starters. I am deeply indebted to all of my students for helping me practice presenting the ideas for this book during lectures, Canvas discussion boards, and informal chats in my office. In particular, I would like to extend a huge thank you to each of these current and former students for their research help: Mark Johnson, Victoria Potts, Molly Winders, Victoria Dowdy, Elane Westfaul, Meg Markin, and Andres Bosque. You were all so wonderful to work with.

I am so grateful to have what are arguably the most fantastic department colleagues that anyone could want. Thank you for your incredible support and friendship. I am also grateful to be part of multiple vibrant scholarly communities, particularly the global environmental politics and feminist international relations communities. It is heartening to know that there are so many dedicated people working with passion on topics that will make the world a better place.

I would also like to thank everyone at MIT Press. I am so appreciative of the opportunity to have my book join so many of my academic heroes in your catalogue. Thank you especially to Beth Clevenger and Anthony Zannino for working so patiently and diligently with me throughout this process.

Finally, thanks to my family. Your love and support mean the world to me. Thanks to Dursun for reading every word of my first draft. Also, while writing this book, I met my favorite person on this planet—my son. While he is still a few years away from being able to read this, I want him to know how grateful I am that he is in my life and how hopeful I am that his generation can overcome the super wicked problems that have stumped ours. Love you, little one.

1 INTRODUCTION: WHAT ARE THE CONNECTIONS BETWEEN GENDER AND CLIMATE CHANGE?

Close your eyes and think of the phrase "climate change." What are the pictures that come to mind? Some might see scientists giving presentations in front of graphs. Some might see communities fleeing floods or wildfires. Some might see people waving signs and chanting at a protest. Who are the people in these pictures? How likely is it that the same person is visualized in each picture—presenting at a conference, walking through a flooded landscape, and attending a protest? Chances are we envision different people in each of these scenes. Why might this be the case? At a very basic level, that is what this book is about. It asks about the roles or characteristics we associate with a certain category of people in climate change debates: women. What characteristics do we expect them to possess? What tasks do we assume they perform? What spaces do we suppose they exist within? And what are the implications of these representations for goals such as gender equity and environmental sustainability?

Numerous environmental problems are urgent and important. The international community faces biodiversity loss, deforestation, water scarcity, and a whole host of other environmental challenges. With all of these pressing environmental concerns, why do we seem to hear so much more about climate change than these other issues? Why is it that climate change seems to dominate the attention of the global media, policymakers, and scholars? According to Sikina Jinnah (2011, 1), "It is difficult to find an international organization, corporation, NGO, university, foundation, religious organization or government agency that does not have a climate-relevant program or focus. . . . It seems that everyone, from McDonald's to the Vatican is jumping on the proverbial climate change bandwagon." Global policy negotiations that at one time would have been focused on a specific environmental issue such as desertification or fisheries are now more often framed through the lens of climate change (Axelrod 2011; Conliffe 2011). For better or worse, climate change has captured our attention in a way that few other environmental issues have.

One major reason for this disproportionate focus on climate change is the urgency and scale of it. Climate change is widely referred to as one of the greatest global challenges faced by the international community. Reports from the Intergovernmental Panel on Climate Change (IPCC) indicate that climate change is currently resulting in a wide range of negative impacts on both humans and ecosystems, and these impacts are predicted to worsen in the future (IPCC 2014a, 2018b). Recent climate-related extremes, such as heat waves, droughts, and floods, reveal substantial vulnerability and exposure to climate variability for many ecosystems and human communities. These kinds of climate-related extremes result in changes to ecosystems, alterations in the availability of resources such as food and water, destruction of infrastructure and damage to settlements, negative impacts on mental health and well-being for many people, and even death. Climate-related hazards serve to exacerbate other stressors, particularly for people in precarious economic or social positions. With the effects of climate change being unevenly distributed, those that are already vulnerable become even more so because of factors such as the uneven distribution of climate change impacts across communities, a deficiency in adaptation capacity, and a lack of access to decision-making structures.[1] Although climate change has broad societal consequences, its impacts tend to be felt most by historically marginalized populations.

From the perspective of a social scientist, one might argue that climate change dominates environmental discussion because it has been politicized in many countries. We see this politicization in disagreements about not only how to address it, but whether it even exists as a problem. In multiple states in the global North, where this book is focused, people on one side of the political spectrum express stronger belief in and concern about climate change than those on the other (McCright, Dunlap, and Marquart-Pyatt

2016; Unsworth and Fielding 2014). In this context, politicians who express support for climate action signal information to voters far beyond their environmental stance. The scholarly literature has attributed this politicization to, among other things, "the behavior of political party elites, the 'anti-environmental countermovement,' and scientists from all sides who engage in public debate" (Pepermans and Maeseele 2016, 479). Topics that are controversial tend to get a great deal of media coverage. There has been a sharp increase in the amount of climate change stories published in the global media since the mid-2000s, particularly across the global North.[2]

Despite the increased media, policy, and scholarly attention to climate change, humanity still has a long way to go before we achieve effective paths to both mitigation and adaptation. This is because climate change is a particularly tricky and complex political problem (Underdal 2017). Kelly Levin, Benjamin Cashore, Steven Bernstein, and Graeme Auld (2012, 124) highlight some of this complexity by conceiving of climate change as a "super wicked problem" with four important characteristics: we are running out of time to effectively stop the problem; the actors who are most responsible for causing the problem are also the ones who are trying to provide a solution; we lack a strong central authority that could help address the problem; and we tend to see policymakers and the public push responses into the future. "Together these features create a tragedy because our governance institutions, and the policies they generate (or fail to generate), largely respond to short-term time horizons even when the catastrophic implications of doing so are far greater than any real or perceived benefits of inaction."[3] Climate change governance has been frustratingly slow and incomplete.

I started graduate school in 2003, the same year that over one thousand people died in heat waves across major European cities. More than a decade later, researchers attributed roughly half of those deaths to climate change (Mitchell et al. 2016). Over the years, climate change has come to dominate several fields of study, including environmental sociology, environmental geography, and my academic home of global environmental politics.[4] Climate change work seems to overshadow our academic journals and conferences (Dauvergne and Clapp 2016). Within academics, climate change research spans the divides of the natural sciences, social sciences, and humanities. Since the 1990s, scholarship has significantly advanced crucial understanding of the scope and scale of humanity's experiences with climate change. Climate change is a difficult political and social problem not because we lack information about it. It is difficult because it is wide-ranging and challenges many current "truths" about how humans can and should engage in economics, politics, and other social processes. In this way, perceptions and understanding of climate change are just as important as the physical processes of rising temperatures, changes to disease vectors, or sea level rise.

Additionally, many different actors with varying interests and perspectives take part in global debates about climate change. Some of these, such as the media and various social movements, have a powerful role in shaping our individual assessments of climate change. Even though media portrayals of climate change differ around the world (Vu, Liu, and Tran 2019), they serve as an influential source of information for average citizens about climate science, policy goals of mitigation and adaptation, and the like (Boykoff and Roberts 2007). Similarly, studies find that environmental nongovernmental organizations (NGOs) can influence understandings of environmental issues (Eden, Donaldson, and Walker 2006; Newell 2000). In these ways, multiple actors shape perceptions of climate change and guide assumptions about the major topics of debate as well as who is having those debates. Thus, while it is essential to understand the physical realities about global issues such as climate change, it is also necessary to make sense of social perceptions of them.

The central contribution of this book is, then, to help fill in some of the gaps in understanding the politics and actions around climate change.[5] It argues that representation matters. The roles and identities assigned to people crucially shape expectations about who should be present in climate change spaces, who is a legitimate voice in those spaces, what jobs and tasks we should assign, and who we should look to for insight about existing problems and paths forward. In particular, this book examines how women are portrayed in climate change debates. It asks questions such as where is women's presence or absence recognized? What tasks are they expected to perform? What factors influence their roles? The book identifies four distinct but related discourses on women and climate change. Each highlights a characteristic, condition, or position associated with women's activities in and experiences of climate change. They provide a multifaceted portrayal of women in/and climate change, but they also demonstrate types of homogenization and generalization that at times can be detrimental to the goals of sustainability and gender justice.

EXAMINING GENDER

Scholars have detailed ways that climate change exacerbates patterns of marginalization and vulnerability as they manifest across different spaces (e.g., the global North versus the global South, rural communities versus urban communities), as well as across different social cleavages such as gender, class, and race (Adger 2006; Barnett, Lambert, and Fry 2008; Bohle, Dowing, and Watts 1994; Brklacich, Chazan, and Bohle 2010; Cutter 2006; Denton 2002; Eakin and Walser 2007). This book touches on all of these divisions but focuses specific attention on gender.

Feminist scholars have been examining the concept of gender for decades. At a very basic level, gender can be viewed as a set of socially constructed ideas about what people identified as "men" and as "women" ought to be and do (Detraz 2017b). Understanding gender to be a product of social construction means that assumptions of "masculine" and "feminine" behavior are not to be taken for granted. There is not a normal or natural way to be, but rather humans are exposed to expectations that (1) they should fit within the category of either "men" or "women" and (2) there are acceptable ways for "men" or "women" to be or act. In the first case, most dominant discourses of gender function as a binary with masculinity on one side and femininity on the other. Individuals are expected to fall on one side or the other, and those who refuse (to) or confuse (us) are regarded as transgressive (Gentry and Sjoberg 2015; Rao 2014). In the second case, while gendered expectations are fluid—they shift over time and across societies—they are still a powerful force for shaping behavior. The existence of multiple forms of masculinity and femininity within a given society at a given time does not blunt the multiple ways that gender works on us every day. Though most can recognize that current understandings of femininity look different from expectations in the early 1900s, for example, this does not mean that we are not still punished or rewarded for complying with dominant expectations of gender today. Additionally, because gender is a social construction, change is possible, but not always easy to come by. The fluid nature of gender results in multiple actors both shaping and reinforcing the dominant discourses about gender. Change, then, requires shifts across a multitude of actors and spaces, many of whom are resistant to alteration.

This view of gender largely speaks to the presence of gender norms in everyday lives, or how gender works on people. Another important aspect of gender is how concepts, discourses, actions, and actors are also influenced by norms about masculinity and femininity and how those associated with masculinity are given greater priority than those associated with femininity. Institutions such as the military, for instance, are inseparable from the norms of masculinity. They guide expectations about the people and actions expected to be present in military and militarized spaces, as well as the behavior and perspective of the institutions themselves (Duncanson 2009). Because of the way gender typically appears in everyday speech, it is often conceptualized as referring only to assumptions about people. This is a limited idea of gender. Actions can also be gendered. For instance, feminist scholars talk about the gendered nature of economic processes (Elias and Roberts 2016; True 2012; Waylen 1997), gendered violence against people and places (Castillo 2008; Staudt, Payan, and Kruszewski 2009; Whetung 2019), and gendered environmental attitudes (Anshelm and Hultman 2014; Pulé and Hultman 2019).

Another shifting aspect of gender is the various ways it relates to other aspects of identity. Gender intersects with race, class, sexuality, caste, ethnicity, and dis/ability in the ways that society understands difference, acceptance, and value. The term "intersectionality" was coined by critical race and legal scholar Kimberlé Crenshaw in the 1980s to highlight the various ways that modes of marginalization interact (Crenshaw 1989, 1994). Over time, intersectional analysis has become widely adopted and adapted across multiple disciplines. Several feminist environmental scholars in particular have argued that intersectionality is an essential component for analyzing how multiple kinds of power differentials work together within environmental issues (Braun 2011; Djoudi et al. 2016; Lykke 2009; Kaijser and

Kronsell 2014; Manning 2016). In addition to paying attention to gender as a multifaceted concept, this book also considers ways that gendered marginalization and agency intersect with race, ethnicity, class, sexuality, national origin, and other forms of "difference." Intersectional research focuses on the simultaneous and interactive effects of these categories of difference. It explores multiple, co-constituted differences.

Understanding gender in the context of climate change discussions involves recognizing how societal ideas of acceptable and appropriate roles and behaviors shape climate change impacts, policymaking, and policy implementation as well as the lived experiences of each of these for human beings. While there are regularized gendered patterns of experience in the realm of environmental change, scholarship has tended not to be examine them closely (MacGregor 2006; Sandilands 1999). Gender "works" on and in society through multiple processes. Examining connections between gender and the environment or gender and environmental action requires thinking through multiple complex assumptions about where environmental damage comes from and how it might be addressed, along with how traits, actions, and people are valued.

GENDER AND CLIMATE CHANGE SCHOLARSHIP

A great deal of scholarship has focused on links between gender and the environment generally and gender and climate change more specifically. Feminist scholars have long explored environmental topics, particularly through work associated with perspectives such as ecofeminism,[6] feminist political ecology, environmentalist feminism, eco-critical feminism, and critical feminist eco-socialism (Detraz 2017b; Lykke 2009; Mies and Shiva 1993; Plumwood 2006; Rocheleau, Thomas-Slayter, and Wangari 1996; Seager 2003; Sturgeon 1997; Warren 1997). While there are important debates that take place among these categories of feminist environmental scholars, one of the threads that tie their work together is their commitment to taking gender seriously in considerations of the causes of environmental change, experiences of environmental degradation, and the economic, social, and political implications of these phenomena.

As discussed above, gender manifests in multiple ways in the daily experiences of individuals. This means that gender works in myriad ways through the causes and experiences of environmental problems along with policy choices of solutions to them. At the same time, gender identities themselves are constructed in part through environmental struggles and practices. Farhana Sultana (2009, 428), for instance, claims that "gendered subjectivities are socially and discursively constructed but also materially constituted; subjectivities are produced through practices and discourses, and involve production of subject-positions (which are usually unstable and shifting). Subjects are always embedded in multiple relations of power, and are interpellated differently across space and time." This means that while socially constructed gender norms influence relationships to our environment, the association also goes the other way: society's very ideas about masculinity and femininity can be bound up with understandings of the environment and our place within it. For example, certain tasks are often considered to be "men's work" or "women's work." An individual's inability to perform those tasks then suggests something about that person's identity as a man or a woman (Nagel 2012).

Feminist scholars highlight the specific associations between the relative position of people in society and the ways that they experience and/or contribute to environmental change. In so doing, they consider the power relations and social norms that shape lived experiences. For instance, some feminist authors claim that the systems of domination that contribute to the marginalization of women and other groups are frequently the same systems of domination that contribute to environmental change (Plumwood 2002; Seager 1999). This kind of work calls attention to the importance of unraveling multiple forms of power relations in order to understand how they influence drivers and experiences of environmental change as well as the policymaking process that address them.

There is a great deal of breadth to the gender and climate change field. Research topics include gendered experiences of climate change impacts, how climate governance is gendered, and how gender shapes understanding of climate change. For instance, several studies examine ways that gender shapes phenomena associated with worsening climate change in the context of a particular state. This work focuses on topics such as gender and adaptation, migration, food insecurity, or natural disasters (Alston 2011; Cannon 2002; Detraz and Windsor 2014; Nagel 2016), often in sites in the global South (Andersen et al. 2017; Azong and Kelso 2021; Haynes 2017; Team and Hassen 2016).

The literature on gender and climate governance highlights both the ways that gender influences policy approaches and participation in climate governance (Kaijser and Kronsell 2014; Magnusdottir and Kronsell 2015; Zhou and Sun 2020). Gunnhildur Lily Magnusdottir and Annica Kronsell (2015), for example, examine the representation of men and women in Scandinavian climate policymaking as well as whether climate policy is gender sensitive. They find that although women were not underrepresented in the political and administrative institutions responsible for climate policy, having a critical mass of women in the institutions has not led to critical acts on gender in policymaking. Gender has not been central to the policymaking process. Studies like these call attention to both gender representation and how gender functions in climate governance.

Another strand of the literature highlights ways that understandings of climate change are influenced by gender norms and gendered experiences (Arora-Jonsson 2011; Dankelman 2010; Djoudi et al. 2016; MacGregor 2010, 2017a). This scholarship explores themes such as gendered discourses or perceptions of climate vulnerability, including ways that women in the global South and global North are cast in predictable roles as either environmental victims or environmental saviors. In the words of Sherilyn MacGregor (2010, 228), "Gender analysis and the study of gender politics should involve the analysis of power relations between men and women and the discursive and social constructions of hegemonic masculinities and femininities that shape the way we interpret, debate, articulate and respond to social/natural/ technological phenomena like war, economic crisis and climate change." Along these lines, some scholars examine gendered climate expertise and the production and reception of climate knowledge by critiquing the supposed value-neutrality of climate science and economics (Nelson 2007; Tuana 2013). While much of this work explores gender broadly or connections between women and climate change more specifically, there is a growing body of academic work on the intersections between men, masculinities, and environmental change as well (Anshelm and Hultman 2014; Enarson and Pease 2016; Kinnvall and Rydstrom 2019; Pulé and Hultman 2021). For instance, scholars examine ways that struggles faced by men and boys during natural disasters can be magnified or exacerbated by assumptions that they are to be stoic and strong (Ariyabandu 2009; Rydstrom 2019; Enarson and Pease 2016). Other studies also trace masculinity to climate denial and fears about loss of social and economic position (Anshelm and Hultman 2014). Martin Hultman (2017) in particular has written extensively on conceptualizing masculinities within environmental discourses. He and Jonas Anshelm remind us that, like the concept of femininities, that of masculinities is "always-in-the-making within and part of material-semiotic antagonistic discourses, which are the embodied nature of knowledge, materiality, meaning and power" (Hultman and Anshelm 2017, 19). Each of these strands of the gender and climate change literature offers important reminders of the numerous ways that understanding and experience of and policymaking and action around climate change is gendered.

There are many different ways to think about gender and climate change. Scholars who contribute to this literature span multiple academic disciplines, and their work is often interdisciplinary. My goal, however, is not necessarily to tell these scholars something dramatically new about gender and climate change. Rather, my goal is to illustrate how people in climate change spaces already use discourses of women and climate change. It is to reflect on how these discourses at times complement existing scholarship as well as often contain storylines at odds with much of the literature. I seek to examine whether and where the discourses are compatible with the goals of environmental sustainability and justice.

In particular, I intend for the book to be of interest and use for scholars in global environmental politics and professionals in other climate change spaces, as well as scholars in women and gender studies. In particular, I seek to shine a light on how many climate scholars and practitioners already think and experience gender without necessarily being aware of it. The discourses presented in the following chapters represent understandings of gender in both conscious and unconscious forms. I argue that these discourses are necessary to evaluate and critically engage with in order to ensure that (1) attention to gender and climate change increases, and (2) this increase takes place in ways that are oriented towards both justice and sustainability goals.

Gender is an underexplored topic within global environmental politics. A quick search for "gender" in the journal *Global Environmental Politics*, arguably the most high-profile journal in the subfield, results in fewer than twenty research articles. There have been attempts to include gender in the field to a greater extent (MacGregor 2017b), but it is a topic that remains underexplored. This is perhaps not that surprising given the history and composition of international relations (IR) and political science more broadly in which both environmental politics and feminist IR have been left out of the core (Green and Hale 2017; Tickner 2014). It is my hope that *Women and Climate Change: Examining Discourses from the Global North* can be part of the conversation about how scholars and practitioners can explicitly think about how gender is already present in our work and what it will look like in a future marked by some form of climate change.

PLAN OF THE BOOK

The following chapters provide a discursive map of some of the qualities, tasks, and positions associated with women in/and climate change. Chapter 2 explains the methodological approach of the book, including some of its similarities to Global Environmental Politics and feminist IR. It provides a detailed account of the feminist constructivist lens used to examine links between power and knowledge that fundamentally shape understanding of gender as well as climate change. This feminist constructivist approach allows for the analysis of discourses that both reinforce and challenge existing gender norms, as well as highlighting multiple forms of agency along with marginalization or exclusion.

Each chapter that covers a discourse begins with an examination of the perspective of an interviewee that illustrates some of the central storylines of the discourse. These are designed to not only highlight the properties of the discourses, but also to offer a glimpse into the world of the women working in climate change with whom I spoke, including the specific roles that women are expected to play or characteristics of women that affect their connections to climate change. Each chapter ends with an evaluation of the implications of the discourse.

Chapter 3 focuses on one of the most visible storylines associated with women: vulnerability and victimhood. These include the idea that women have socially mandated responsibilities that intersect with climate change impacts, and that they often lack adaptative capacity to effectively deal with climate change. It sees women as uniquely burdened by climate change. It explains how interview participants utilize this discourse in ways that both conform to typical discussions of vulnerability in environmental politics and policymaking as well as reject the idea that gender necessarily connects to the role of victim.

Chapter 4 addresses connections between women, caregiving, and climate change. It explores the role of the caregiver as it relates to climate change causes, impacts, and policies. While these care storylines are often connected to women through a specific focus on mothering, they also relate to other socially conditioned care expectations that are regularly associated with women and femininity. The chapter specifically highlights storylines such as caregiving roles influencing women's thinking about climate change, motivating them to get involved in climate change action, and shaping their approach to parenting. It also details ways in which actors use care and motherhood storylines strategically in climate change work.

Chapter 5 covers the discourse of women as knowledgeable about climate change and considers how women tend to be less likely to engage in climate skepticism and more likely to educate others about climate change. The chapter also features an examination of some of the roles that participants identified for knowledgeable women. These roles include climate change leader/diplomat, someone who provides knowledge to others, and someone who generates climate change information through research.

Chapter 6 examines a range of roles that women are assumed to play in the realm of climate change. This includes potentially positive as well as negative roles related to environmental sustainability. Thus, the chapter explores how routine activities that women engage in can drive climate change and but also how they deploy a range of strategies to combat climate change.

Additionally, the chapter focuses on ways that women's participation in climate change work is denied, overlooked, or undervalued. It highlights multiple storylines related to the idea of women being constrained, limited, or underestimated, ranging from women's literal absence in some climate change spaces to ways that their contributions are belittled or undervalued. Finally, the chapter examines some of the consequences of action that interviewees described for themselves and others.

The final chapter assesses the four discourses taken together. The chapter offers suggestions about the implications of the dominant discourses and storylines, as well as reflections on how we as scholars might learn from these depictions of women and climate change in our quest for the transformations required by environmental sustainability. It puts forward frames of representation that allow us to be more inclusive and reflective of women and the range of tasks they currently perform. Taken together, the chapters shed light on the multifaceted discourses used to describe and comprehend women and climate change. They illustrate ways that our stories of women and climate change conform to stereotypical understandings of gender and how they challenge them. They reveal existing limitations as well as the potential for developing complex frames of women and climate change that are useful for shifting our debate into new terrain.

2 EMPLOYING A FEMINIST METHODOLOGY

Between 1997 and 1998, a debate about methodology in international relations took place within the journal *International Studies Quarterly*. In a prominent exchange between Ann Tickner (1997) and Robert Keohane (1998), Tickner noted the frustration that many feminist scholars feel when speaking to an audience unfamiliar with their approach.[1] Rather than engaging with the substance of their argument, the audience often ask questions that seem to challenge whether they are actually doing work that fits within the discipline at all. Brooke Ackerly (2009, 432) characterizes the Tickner-Keohane debate as a "normative feminist theorist asks a positive theorist to reconceive the conceptual building blocks of the IR [international relations] field and a positive theorist asks a normative theorist for a positive research agenda." For many years I have assigned these two articles to students in my Scope and Methods of Political Science graduate seminar. The response is often something along the lines of "they don't really seem to be talking to each other!" Students note that both scholars appear to have something very different in mind when they talk about international relations—its characteristics, goals, and expectations. While there are frequent calls for methodological plurality in IR, it remains the case that many scholars within the field remain unfamiliar with the kinds of methodologies associated with feminist scholarship. Since one of the key goals of this book is to reinforce the importance of centering gender in global environmental politics (GEP), it is important to illustrate how this is done with a specific feminist methodology. Additionally, focusing on methodology allows us to see an important area of similarity between some GEP scholars and feminist scholars.

This book sets out to highlight the existence of discourses on women and climate change that are already being utilized by people in climate spaces. I argue that gender is of fundamental importance for understanding environmental politics, including the politics of climate change. Because of this, a feminist methodology has guided my work from the initial stages of pondering my research puzzle to the final stages of editing the manuscript. Feminist approaches to international relations gained prominence in the decades after the end of the Cold War. There is not a single feminist approach or methodology, but rather significant diversity among feminisms (Ackerly, Stern, and True 2006; Ackerly and True 2010). The majority of these fall within the category of critical theories rather than problem-solving ones in the conceptualization of Robert Cox (1981, 129). According to Cox, critical theory "stands apart from the prevailing order of the world and asks how that order came about. Critical theory . . . does not take institutions and social and power relations for granted but calls them into question by concerning itself with their origins and how and whether they might be in the process of changing."[2] Meryl Kenny and Fiona Mackay (2018, 92) observe that "feminist approaches are explicitly political in that they seek not only to recognise and understand gender power relations, but also to change and transform them." Likewise, Brooke Ackerly (2009, 433) calls on us to think of feminism as a research ethic or "a theoretically informed practice that affects every decision about research, from theoretical predisposition to question to method of data collection to method of analysis to publication." This necessitates "feminist rigor in research, a rigor directed at reflection on all stages of the methodology." While many scholars consider gender as a variable within international relations, what makes work explicitly feminist is its commitment to a particular methodology, or system of methods (Wibben 2016). In the case of feminist environmental scholarship, it centers a normative commitment to justice and equity as well as sustainability. None of this is to suggest that we should be content with scholarship that is not rigorous, but rather that the criteria we use to evaluate methods take into account multiple ways of doing good work.

Cynthia Enloe (2016, 258) argues that "a feminist approach to anything . . . is an approach that is rooted in awareness that the researcher is part of a collective. That consciousness translates any of our efforts into engagement with the work of others, others to whom we are accountable. No matter how alone any feminist might occasionally feel when conducting research or writing, we can never kid ourselves into imagining that we are living on an intellectual island." In the case of my scholarship, I see myself embedded in multiple academic fields including global environmental politics, feminist international relations, and feminist environmentalism. It is essential to recognize these communities, particularly given my overarching goal of highlighting discourses of gender within climate change. My scholarship has been influenced not only by my own ontological and epistemological position but also these intellectual communities.

Like feminist IR, GEP has often utilized methodological approaches that separate it from the mainstream of political science. This is not to say that there is no diversity within the field of GEP. One aspect of this diversity is a split along epistemological positions with some scholars aiming for positivist causal explanations and others rejecting, in whole or in part, the idea that we can provide this kind of "knowledge." Over the years, GEP scholars have adopted numerous methodological approaches across this divide (Hochstetler and Laituri 2014). Likewise, within the community of GEP scholars there has tended to be an openness to acknowledging that multiple kinds of methods can result in rigorous academic work (Neville and Hoffman 2018). While the majority of GEP scholarship has tended to use qualitative methods over the years, more recently there has been an increase in the use of quantitative methods. However, not everyone finds these shifts in methodology to be necessarily positive. Peter Dauvergne and Jennifer Clapp (2016, 3) claim that "the increasing complexity of theories and intricacy of modeling and statistical methodologies risk disconnecting contemporary scholarship from the earlier goals of problem-focused, policy-oriented, activism-linked research." This suggests that one's methodological approach should fit the goals of one's scholarship rather than being driven by the conventions in one's field of study.

Some argue that certain methods or approaches should be chosen because of the unique characteristics of environmental politics topics. For instance, Kate O'Neill and coauthors (2013, 441) state that global environmental governance problems are often "characterized by institutional and issue complexity, linkages, and multiscalarity that pose challenges for many conventional methodological approaches. As a result, given the large methodological toolbox available to applied researchers, we recommend they adopt a reflective, pluralist, and often collaborative approach when choosing methods appropriate to these challenges."[3] This argument suggests that there is something unique about the kinds of challenges or concerns that drive much GEP scholarship and that these should influence our methodology.[4] They go on to argue that methodological choices should be made based on their utility to explain and understand "the questions that motivate our research—usually to address an environmental problem—rather than choosing methods first and then selecting research questions those methods can address" (O'Neill et al. 2013, 442). While these calls for critically evaluating methodology are sound, adopting a feminist methodology implies that one's ontological and epistemological positions are inseparable from one's research choices. In the words of David Marsh, Selen A. Ercan, and Paul Furlong (2018, 177), they are a skin not a sweater. They are something that we live in rather than something that can be shed at will only to be replaced with something else for the next project. Feminist scholars are guided by their view of how they see the international system, and this will influence the questions they choose to ask about it as well as their approach to scholarship. Feminist methodology relates not only to how scholars design their research, but also to their larger goals for scholarship, which are explicitly normative.

Likewise, many if not most GEP scholars ask the questions they ask because they are motivated to do more than simply test theory. They want to know about the world because they seek particular kinds of changes within it—sustainability or justice or equity. Kate Neville and Matthew Hoffman (2018) have argued that GEP scholars should be up front about our normative position both for greater transparency in research and in order to foster trust between academics and the general public. Feminist IR and feminist environmental scholarship illustrate this kind of openness about their normative commitment (MacGregor 2017b). Those who make up this academic community—across disciplinary homes—are motivated by a deep concern about not only gender justice, but also the challenges facing the planet. This kind of transparency about goals allows us to center justice in approaches to studying climate change.

This book undertakes a critical analysis of climate change discourses that feature a focus on women. It highlights the characteristics associated with women in these discourses, and the roles in which they are cast. In order to better understand the discourses of women and climate change, I chose to speak to women actively working in climate spaces. Qualitative interviews are an appropriate choice because I see knowledge as "something which does not already exist, but which is created and negotiated in the interview, with both interviewee and researcher actively participating and interpreting" (Yeo et al. 2014, 179). This approach to interviewing understands the process as potentially transformative for both parties and sees the researcher as an active participant in the development of data and meaning (Kvale and Brinkmann 2009).

The data collected for this project can shed light on the perspectives of women on the question of whether and how gender influences the causes of, experiences of, and solutions to climate change. These first-hand accounts from women with expertise in a variety of climate change areas represent an important contribution to the way we understand climate change as a gendered phenomenon and the discourses we use when describing women's place in climate change debates. Discourses refer to the way we make sense of the world. They are powerful forces within both academic and policy debates (Milliken 1999). I am guided by Maarten Hajer's (1995, 45) definition of discourses as "specific ensembles of ideas, concepts and categorization that are produced, reproduced and transformed in a particular set of practices and through which meaning is given to physical and social realities." This definition suggests that discourses are constantly evolving entities that are shaped by society over time. At the same time, actors can draw on discourses strategically in order to focus attention on a particular issue or frame an issue in a specific way. The process of discourse analysis involves tracking the storylines that make up a larger discourse. A storyline is a set of concepts, ideas, or themes that are repeated and combine to form a discourse. Each of the discourses outlined in the following chapters focuses on particular elements of women's characteristics or roles in climate change. They are made up of multiple storylines. According to Fran Tonkiss (2004, 373) "discourse analysis involves a perspective on language that sees this not as reflecting reality in a transparent or straightforward way, but as constructing and organizing terms in which we understand that social reality. Discourse analysts are interested in language and texts as sites in which social meanings are formed and reproduced, social identities are shaped, and social facts are secured." This is consistent with an anti-foundationalist ontology that is common among multiple approaches to IR, including some versions of feminism, constructivism, and poststructuralism. Within this position "the material world is not necessarily rationally and objectively independent of our conceptualization of it because we mediate reality through our interpretations and emphases. Discourse is the mode of interpretation" (Gentry 2016a, 24).

Thinking through meaning and discourse is essential in climate change work because discourses can guide debate and policymaking in distinct ways. For example, Maria Stern (2006, 181) defines discourse as "the production and re-presentation of meaning, which delimit the realm of understanding, action, and imagination within a certain framework." There is not one dominant discourse used to frame women's position in climate change. Rather, there are multiple, overlapping, and sometimes conflicting discourses. This multiplicity is important, because it contradicts the frequent tendency to paint women as uniformly vulnerable across the international community. According to Annick Wibben (2011, 4), "there is always more than one point of view and more than one story to be told. The choice to privilege one perspective over another is never innocent or obvious but always intensely political." The fact that vulnerability discourses have had an oversized place in climate change debates illustrates that one particular representation of women, and particularly women from the global South, has disproportionately tended to shape our imagination on climate change (Arora-Jonsson 2011; Cuomo 2011; Denton 2002; MacGregor 2009). In line with work by Sherilyn MacGregor (2010) and Mary E. Pettenger (2007), I use a feminist constructivist lens to examine links between power and knowledge that fundamentally shape understanding of gender as well as climate change. This feminist constructivist approach allows for the analysis of discourses that reinforce existing gender norms as well as challenge them. The approach is particularly helpful for highlighting multiple forms of agency in addition to pointing out marginalization or exclusion. This kind of analysis is essential for determining not only how we as scholars or practitioners might address climate change more effectively, but also for how we might overcome rigid gender roles and intersectional marginalization as we envision alternative futures that become necessary due to environmental change.

This book is built from the analysis of semi-structured interviews with seventy-six individuals who identify as women working in climate change spaces in the global North in different capacities, including as academics, nonprofit workers, activists, and practitioners.[5] Interviews were conducted between August 2016 and June 2017 and in March 2019. As shown in the appendix, participants were individuals who self-identified as women and who engaged in some kind of climate change-related activity at the time of the interview. The majority of the interviewees held jobs working for environmental nonprofits or were scholars working on some facet of climate change. A few participants were either interns or volunteers for nonprofits. Most of them identified themselves as white and financially stable. Very few identified themselves as women of color. While I tried to increase the number of non-white respondents in the study, this was a difficult task for a few reasons. First, academics as well as the environmental nonprofit sector tend to be majority white spaces (Cimpian and Leslie 2017; Montañez 2018; Taylor 2018). Second, while there is a strong representation of women of color in environmental advocacy—particularly in organizations campaigning for environmental justice—they are often overburdened with service and would be less likely to have the time to speak to me.[6] I used purposive sampling to identify participants who had experience with climate change debates but might not have specifically thought a great deal about gender and climate change (Ackerly and True 2010). I used a snowball technique (i.e., asking interviewees to recommend other candidates for participation) to expand the pool of candidates.[7]

I conduced thirty-four in-person interviews, twenty-five through a program such as Skype, Google hangouts, or Zoom, and seventeen over the phone. I started with the same list of questions for each interview, then asked follow-up questions when necessary to draw out a point, encourage a participant to elaborate, or clarify a comment for my own understanding. Each interview typically lasted between thirty minutes to an hour. The interviews were all recorded and then transcribed into a text file.[8] I maintained the anonymity of the participants by assigning them pseudonyms. While there are arguments to be made both for having participants remain anonymous and for using their names, I chose to keep interviewees anonymous because I wanted them to feel comfortable speaking freely about their own actions in climate change as well as their institutions or organizations, and I felt that they would feel more comfortable doing this if I used pseudonyms. Additionally, I sent out a survey containing the interview questions in order to serve as a check on the results that I got through the interviews. I sent out the survey link through two listservs that I am a part of. One of them is on environmental politics and the other one is on feminist theory and gender studies. I had twenty survey responses.[9]

The interview and survey transcriptions were coded for discourses of women's roles and characteristics in climate change. This involved looking for themes, terms, and ideas that indicate a link between women and climate change. These themes, terms, and ideas were developed inductively through coding the data multiple times. I coded the data first by hand, then in NVivo software (Coffey and Atkinson 1996). The analysis revealed four central, overlapping discourses describing connections between women and climate change.[10] Each of these central discourses is made up of multiple storylines. In general, the responses to the survey questions followed patterns that were similar to those of the interviews, with the same central discourses used.

Interviews offer a rich source of information for a project such as this. However, it is essential to be clear about the position that the interviewees were put in by agreeing to take part in the project. They had no prior information about the questions that I would ask. Most of them had no background in thinking systematically about connections between gender and climate change. Their experience and expertise are in climate change, and many of them had not thoroughly thought about gender and climate change before I sent them an email asking them to participate. Essentially, they were put on the spot to offer reflections on extremely complicated and fraught issues. Several participants were noticeably uncomfortable about being in this position. They mention things like "I struggle to not be offensive, or incorrect in my views," or "I'm hesitating slightly because what I'm uncomfortable with is a broader stereotype." This means that many of them were drawing on existing discourses or impressions they had readily at hand. Having participants answer questions without advanced preparation is useful because it gives a better sense of which discourses are dominant among people who have spent a great deal of time thinking about climate change, but not how climate change intersects with gender. Charlotte Epstein (2008, 7) argues that discourses allow individuals to "make sense of themselves, of their interests and their ways of behaving, and of the world around them." The discourses and storylines that the interviewees used reflect not only their understanding of themselves and their position within climate change, but also their perception of larger human-climate change interactions.

WHY FOCUS ON WOMEN?

A thorough examination of climate change necessitates a nuanced, critical evaluation of how power and perception shape life. Since climate change is a global phenomenon that touches nearly all humans on the planet, some may wonder why a specific emphasis on perceptions of the characteristics and roles of women is necessary. I chose to focus on women in climate change debates for a few reasons. First, because the varied and complex experiences of women around the world have frequently been homogenized into simplistic narratives that diminish women's agency. The project seeks to illustrate the hazards of painting the world's women with a broad brush. Assuming homogeneity erases differences in how climate change is experienced, studied, and addressed. To identify effective and just solutions to climate change, it is essential to have a clear picture of the various roles women already play and might play in combating climate change now and in the future.

Second, I focus on women because they remain invisible or underestimated in many aspects of climate change debates. In 1990, pioneering feminist international relations scholar Cynthia Enloe (1990, 1) argued that "if we employ only the conventional, ungendered compass to chart international politics, we are likely to end up mapping a landscape peopled only by men, mostly elite men. The real landscape of international politics is less exclusively male." Much the same can be said about the realm of climate change. While there has been an increase in prominent women in climate change spaces, most high-profile figures who do the debating are still men, on the side of those who call for urgent action on climate change and of those who doubt its very existence. Specifically, most climate negotiators as well as most policymakers who create and implement policy about climate change are men, and even most newscasters who discuss climate change are men (Ivanova 2015; MacGregor 2009). The perceptions of men are already fairly visible. As in many other topics, men's voices are abundant. And while women are active in most areas of climate change scholarship, advocacy, and policymaking (a point that will be made across the following chapters), these contributions are often not recognized or valued in the same way as men's. This is particularly true for Black, Indigenous, and People of Color (BIPOC) women.

Finally, it is also important to examine the place of women in climate change debates because numerous actors in the climate change arena have already called on us to take gender seriously. Some have highlighted the fact that women tend to be disproportionately burdened by climate change (Dankelman 2002; 2010). In countries around the world, women fall disproportionately into the category of marginalized due to dominant ideas about the appropriate and acceptable roles that men and women ought to play in society (Enloe 2004; Sjoberg 2010; Tickner 2014; True 2012). While it is dangerous and unhelpful to simplistically paint women as "victims" of climate change, a topic that will be discussed in chapter 3, it remains true that gendered patterns of marginalization (intersecting with race, ethnicity, sexuality, dis/ability, class, caste, and so on) mean that many women will be some of the worst sufferers of the destructive impacts of climate change. On the other hand, women have been put forward as a great hope for combating or adapting to climate change. According to Lakshmi Puri (2015), the Deputy Executive Director of UN Women:

> We have to recognize the power of parity and we have to do it now because we can no longer afford to dismiss and waste the potential of women's agency and their huge role in devising and leading responses to climate impacts. . . . The force multiplier and transformative potential of empowered women and girls

should be harnessed to arrest and reverse climate change and adapt sustainably to its impact.

This passage indicates the view that women and girls have a uniquely essential role in humanity's current climate challenge. Thus, women have been cast in multiple roles in climate change theorizing and policymaking. These include victims, drivers of change, laborers, and saviors (Arora-Jonsson 2011; Awumbila and Momsen 1995).

For these reasons, women cannot be understood as a homogenous category. Feminist scholars writing about gender in general, and women specifically, have to walk a fine line between avoiding essentialization, while identifying regularized patterns of behavior and lived experiences. Likewise, the decision to focus on the perspectives of women for this book does not imply that the crucial ways that masculinity and climate change intersect should be ignored. As argued above, there has been important research on these connections. A specific focus on discourses of women will necessarily include reflections on masculinity and men. Simply put, all of these concepts are fundamentally intertwined in the ways that gender works.

WHY FOCUS ON WOMEN WORKING IN THE GLOBAL NORTH?

The women I spoke with were all based in the United States, Canada, Germany, England, and Scotland. Some of them were originally from countries in the global South (Bangladesh, Brazil, Morocco, Taiwan, and Venezuela) or have lived or conducted research in the global South,[11] but the majority were both from the global North and working on climate change issues within the context of the global North. Their position affects how they view climate change, as well as their likely agency in addressing climate change and their vulnerability to its impacts. Unfortunately, no book can cover all the complexity of gender and climate change, or even representations of women in climate change. I chose to focus on the perspectives of these women in the global North because a great deal of scholarly and policy debates about both gender and climate change are dominated by discourses and voices from this part of the world. This makes it an essential space for exploring how gender is understood to work in environmentalism, policymaking, economics, and other areas. Dominant global discourses about gender have tended to come out of the global North and travel to the global South through processes like globalization and imperialism (Mohanty 2003a, 2003b). It is important to understand how people in elite spaces of the global North conceptualize gender and climate change connections because these perceptions will likely be those that influence international climate negotiations, the strategies of multinational corporations, and the environmental policies of northern states with large greenhouse gas emissions levels.

I also chose to focus on women working in the global North because the majority of existing work on gender and climate change is about women in the global South, in countries such as Bangladesh (Alston 2015; Cannon 2002; Haynes 2017), China (Moriggi 2017), Ethiopia (Team and Hassen 2016), Nepal (Bhattarai, Beilin, and Ford 2015; Oven et al. 2019), Nicaragua (Gonda 2017), and Nigeria (Akinsemolu and Obafemi 2020), and many others. This important work makes an essential contribution to our understanding of these connections. At the same time, there is something to be said for also including a different perspective: that of women who have lived experiences and perceptions about climate change from the perspective of the global North. It makes little sense to strive for inclusive, holistic thinking about climate change and how to address it and not include as wide a swath of perspectives as possible. It would not make any sense to reflect *only* on the positions of women in the global North, but it would also not do to ignore them. To date, there is much less scholarly treatment of this perspective. Existing work that focuses on the global North has tended to examine European countries, although some does examine Australia and other northern states (Alston 2011; Cohen 2017; Dymén and Langlais 2017; Fuchs et al. 2017; Magnusdottir and Kronsell 2015, 2021). This book argues that these works on the global North are a complement to the previous research that has already been done, and continues to be done, about the global South.

In addition, I choose to interview women living and working in the global North because I wanted to understand their views on women's vulnerability to climate change. It is well known that vulnerability is a spectrum and that some people who are more vulnerable to the effects of climate change. Most of the women in my interview pool were not on the most vulnerable end of the spectrum. This means that they were not likely to be drawing on personal experiences of direct vulnerability, even as they did use victimhood and vulnerability discourses to describe women's position in climate change. Their doing so presents a unique challenge. How do we avoid talking about women as victims if so many of us engage in it? I was motivated to write on this topic because the discourses describing women in climate change at the global level tended to focus heavily on their victimhood. When I did a quick Google image search for the phrase "women and climate change" the first several pictures that came up were of women presumably from the global South, standing outside, performing some kind of agricultural work or water collection. I had to go down to the tenth image before I saw any variation in this theme; the tenth was of women at a protest for climate justice. I wanted to hear from women who are knowledgeable about climate change and its effects, but who are not necessarily on the front lines of climate change in the way that other communities might be. If they also use victimhood discourses to describe women's position, what does this mean for climate change debates moving forward? What I found was that while participants categorized women beyond the simplistic women-as-climate-change-victims trope, they still tended to stereotype women, particularly women from the global South.

While the perspectives of seventy-six women working in the global North are not generalizable to women as a whole, these perceptions do offer important insight into how climate change is understood in countries that are frequently depicted as leaders in climate change knowledge or action. In 2015 the Pew Research Center observed a gender gap in levels of concern about climate change in several countries in the global North, including the United States, Canada, Australia, the United Kingdom, South Korea, Germany, and Spain. In each country, women were more likely than men to identify climate change as a serious problem. They were also more concerned that it will harm them personally and that and major lifestyle changes are needed to solve the problem (Zainulbhai 2015). This means that even if gendered patterns of marginalization in the global North look different from those in the global South, it is likely that there are still gendered patterns of experiences and perceptions that should be considered in climate change scholarship and policymaking.

What can be learned from the perspectives of northern women without a strong background in gender studies, feminist international relations, or other related fields? These are women who are actively thinking about climate change, and the discourses they use are important for understanding how gender norms are (re)produced or contested in climate change debates. They are also important because they can help us understand where points of contestation lie and what space might be available to transform existing discourses to better achieve goals of sustainability and justice. Interviewees play multiple roles in climate change: some are climate scientists who testify to Congress or work for an environmental nonprofit; others are social scientists who are knowledge brokers or who advise on policy. Critically reflecting on discourses of women and climate change allows us to think not only about gendered patterns of vulnerability, caregiving, knowledge creation, and agency, but beyond these as well. The four discourses outlined in the following chapters often feature storylines associated with dominant gender norms. There is evidence of participants using gender stereotypes in ways that reinforce problematic patterns of "doing" environmental politics, but there are also storylines that complicate or reject rigid gender norms. Thinking through these discourses encourages thinking about how climate change is understood, how narrowly or broadly expertise is conceptualized, and how everyday environmental experiences are gendered as well as raced, classed, and placed.

3 WOMEN AS VULNERABLE: PRECARIOUSNESS IN THE FACE OF CLIMATE CHANGE

Farhana works for an Islamic international aid agency in London. Many of the tasks in her position are directly related to climate change. Her family is originally from Bangladesh, and she started our conversation by telling me that she thinks a great deal about gender and climate change in the context of her organization's environmental work in that country. She went to Bangladesh in 2014 to work in areas that were still recovering from Cyclone Aila, a massive storm that hit multiple South Asian countries in 2009. Farhana noted that "since seeing the disastrous effects of climate change on particularly vulnerable women and children in the communities that flood, I have been passionately campaigning and doing advocacy work for people to invest in climate mitigation and adaptation." Her experience with precariously positioned populations inspired her to focus on climate change work more broadly.[1] In the course of our conversation, she emphasized women's vulnerability to climate change, calling attention to food insecurity, livelihood insecurity, care responsibilities, and the intersection of existing marginalization and climate change adaptation capacity. When I asked for her first thoughts about the phrase "gender and climate change," she responded:

> I particularly think of vulnerable women who are living in areas that are most affected by disasters and I also think about how a lot of the women that I have met . . . are struggling to grow their own crops and food, and are struggling to, for example, have safe housing. Some of them are forcibly having to migrate to areas that are, I guess, less prone to disasters—even though Bangladesh is one of the most prone to disasters and flooding in the world. I also think about

> how those women have struggled to provide for their children. Essentially because they don't have access to growing their own crops unless they have that intervention from NGOs like ourselves. . . . I've found that, definitely, the communities that I visited, there is a lack of literacy rates in some of those remote and rural parts of Bangladesh and to be able to educate those women and children is probably quite lifesaving.

Her response outlined several of the ways that women in Bangladesh are forced to cope with climate change impacts under already constrained conditions, including coping with food insecurity, insecurity of housing, pressure to migrate, inability to care for families, and lack of education.

Farhana was by no means alone in identifying women as being uniquely burdened by the impacts of climate change. In this chapter, I call this set of ideas a *women-as-vulnerable* discourse. In it women tend to face socioeconomic and political discrimination in society, and this leads to them being disproportionately affected by climate change. Storylines related to this include the idea that women have socially mandated responsibilities intersecting with climate change impacts, and they often lack adaptative capacity to effectively deal with climate change. Marginalization leads to women feeling climate change "first and worst." In the words of Allison, a US-based academic,

> Women are in the bottom rung of the social structure in a lot of places. And so . . . as climate change makes it harder to do agriculture, makes it so that there are more floods, and more sea level rise, and all that stuff, I think that women will probably get the short end of the stick. Particularly in places where they already get the short end, it's going to be a shorter end.

In this discourse, it is the intersection of women's position in society with the adverse effects of climate change that leads to their vulnerability.

The following sections assess some of the most frequently mentioned ways interviewees said women are likely to "feel" climate change acutely. They discussed connections between vulnerability and divisions of labor as well as experiences during natural disasters. Additionally, the chapter explains the participants' focus on existing structures as facilitating exclusion, marginalization, and climate vulnerability. We then move on to a discussion of specific categories of marginalization that are present across the interviews. Finally, the chapter suggests a few major reasons for why the women-as-vulnerable discourse might be relatively prevalent in climate change discussions and some implications of using victimhood discourses in general.

VULNERABILITY AND CLIMATE CHANGE

Vulnerability is a central concept for environmental scholarship and policy discussions. Scholars identify various entities that are vulnerable to climate change, including species (Pacifici et al. 2015), states (Kim and Wolinsky-Nahmias 2014), and individuals or groups (Adger, Eakin, and Winkels 2009; Bohle, Downing, and Watts 1994; Kelly and Adger 2000; Gaillard 2010).[2] The existing literature identifies indigenous groups (Shearer 2012), those who experience poverty (Formetta and Feyen 2019; Leichenko 2014), underrepresented communities (Shepherd and KC 2015), and women (Azong and Kelso 2021; Denton 2002), among others, as being particularly vulnerable. Those concerned with social or human vulnerability to climate change outline the multifaceted and fluid factors that influence how individuals or communities are susceptible to climate change. Many frameworks view vulnerability as having both bio-physical and human factors, meaning that it goes beyond simple physical exposure to particular conditions (Barnett 2020). Those who are vulnerable tend to be the ones who also experience some form of marginalization in society. For this reason, most conceptualizations of vulnerable people involve not only an increased likelihood of their experiencing the negative impacts of climate change, but also a reduced ability to cope with those impacts (Kelly and Adger 2000). This capacity-based component speaks to the ways marginalization influences some people's ability to address their condition. For instance, imagine two small-scale farmers who are trying to get through a period of drought. The first reacts by drawing on savings and looking to other sources of income as coping mechanisms. The second farmer has very little savings or alternative sources of income and is thus susceptible to the impacts of the drought in a way that the first farmer is not. This aspect of vulnerability requires us to think beyond the immediate, physical experiences of environmental change to consider issues such as power relations and social embeddedness (Ribot 2010).

Vulnerability is also frequently discussed by feminist scholars, although it is a heavily debated concept in these circles. There has been a great deal of feminist theorizing about whether the label "vulnerable" might serve to disempower those to whom it is applied and how identifying vulnerability might indicate corresponding obligations for states and other actors (Butler 2006; Gentry 2016b; Mackenzie, Rogers, and Dodds 2014). Since the term "vulnerability" originates from the Latin *vulnus*, meaning wound, it calls to mind suffering and fragility, which are part of the human condition. As human beings with fragile bodies, we are prone to suffering physical harms. According to Bryan Turner (2006, 29), one cannot understand vulnerability without reflecting on the fact that we have "an organic propensity to disease and sickness, that death and dying are inescapable, and that aging bodies are subject to impairment and disability." While all human bodies are vulnerable, our placement in social, economic, and political structures dictate which of us are more susceptible to harm than others. Feminist environmental scholars such as Susan Buckingham (2020, 65) argue that "our bodies can only be understood in relation to their environments, both of which are produced by the interaction between social and physical processes." Likewise, Iris Marion Young (2002) and Toril Moi (2001) use the idea of "lived bodies" to highlight how gendered embodied selves exist within specific environments. Gender as well as race and class norms mean that women tend to have reduced access to education, livelihood security, political power, among other things, which in turn increases their likelihood of impoverishment and vulnerability (Elborgh-Woytek et al. 2013).

Some existing scholarly work that evaluates gender and climate vulnerability in a specific place highlights how gender norms both vary and how they are similar around the world (Andersen, Verner, and Wiebelt 2017; Huynh and Resurrección 2014; Furusa and Furusa 2014). This work also often stresses that gender works in complex ways. For instance, in a study of the implications of climate-related water scarcity for women in Central Vietnam, Huynh and Resurrección (2014, 226) found that vulnerability to water scarcity is shaped by multiple factors. The study stresses "the heterogeneity of women as a group and their intersectional dynamics as they adapt to increasing agricultural water scarcity on their rural livelihoods. The findings show that social differences including gender, class, household headship, age and stage of life shape women's differentiated experiences in vulnerability in access to water, to forestland and credit; [these] in turn mark their adaptation differentiation to climate-related agricultural water scarcity." Additionally, a 2017 study across Brazil, Mexico, and Peru found that contrary to the expectations of a simplistic women-as-vulnerable assumption, female-headed households tended to be slightly less vulnerable and more resilient than male-headed households. While the study distinguishes between different types of female-headed and male-headed households and recognizes that there are important variables that influence vulnerability as well as resilience, it does serve as a reminder that assumptions about vulnerability are not always empirically accurate and that careful assessment of the sources of vulnerability are necessary (Andersen et al. 2017).[3] Taken together, this work signals that simplistic portrayals of gendered vulnerability are often partial if not incorrect.

Many scholars have issued strong cautions against treating women as a homogenous vulnerable group in environmental debates (Arora-Jonsson 2011; Cuomo 2011; Denton 2002; MacGregor 2009). They argue that simplistic representations of vulnerability and environmental change both serve to reinforce damaging gender stereotypes and lead to ineffective policymaking—points that will be taken up later in the chapter. Scholars such as these urge us to think critically about the connections between gender and vulnerability in order to identify the processes that lead to some people being more susceptible to environmental harm and less able to address it (Alaimo 2009). This means taking account of multiple forms of marginalization and how they intersect with climate vulnerability (Cuomo 2011).

In addition to work that assesses how gender shapes experiences of environmental vulnerability, scholars examine how gender affects evaluations or assessments of vulnerability (Arora-Jonsson 2011; McCright and Dunlap 2011; Norgaard and York 2005). For instance, some in the fields of social psychology and environmental sociology explore how gendered environmental attitudes could be linked to gendered perceptions of risk

(McCright and Dunlap 2011, Xiao and McCright 2014). According to the "vulnerability thesis," white men feel less vulnerable to many risks than do women and people of color and are more accepting of such risks. This has been termed the "white male effect," which is thought to arise due to their dominant position in social structures (McCright and Dunlap 2011). This work indicates that gender norms and roles, alongside race, class, and other factors, influence understanding of environmental change and how likely we are to see ourselves as vulnerable. Research has found a consistent, albeit relatively modest, connection between gender and concern over or perceived seriousness of different types of environmental problems (Davidson and Freudenburg 1996; Mohai 1997; Sundström and McCright 2014; Xiao and McCright 2014), conceptualizations of environmental and economic trade-offs (McStay and Dunlap 1983), pro-environmental attitudes (Stern, Dietz, and Kalof 1993), and involvement in pro-environmental activities (Hunter, Hatch, and Johnson 2004). Much of this scholarship is focused on the global North. For instance, Aaron McCright (2010) shows that women in the United States express slightly greater concern about climate change than do men. Similarly, Chenyang Xiao and Aaron McCright (2014) find that women in the United Sates are more concerned than are men about health-related environmental problems. Taken together, the scholarly literature on gender, environmental change, and vulnerability illustrates a complex picture of the sources of vulnerability, how members of society experience vulnerability, and how individuals and scholars understand both of these phenomena.

FEELING CLIMATE CHANGE

Much like the scholars and policymakers who use the concept of vulnerability to understand how climate change is experienced throughout society, interview participants often used a vulnerability discourse when reflecting on whether and how gender and climate change are related. Most interviewees who used the women-as-vulnerable discourse referred to the social construction of gender. Women were seen as vulnerable because of the socially conditioned expectations regarding the appropriate or acceptable roles they should play. Most avoided depicting women as vulnerable because of some natural or unchangeable condition. Rather, they explained that women likely feel climate change impacts first and worst because of their roles in families and communities. For instance, a majority of participants mentioned women's labor intersecting with climate change impacts at least once in the course of our discussion.[4] In fact, this is the most frequently used storyline in the women-as-vulnerable discourse. Examples include women collecting water or wood/fuel for the household, women cooking or providing food, and women involved in agriculture.[5] Belina, a Brazilian environmental nonprofit worker living in England, brings up some of these tasks:

> In developing countries, not all of them, not all at the same time, but women tend to, in very poor countries, they tend to be the ones who are in charge of fetching water, fetching firewood, finding something to eat for dinner, cultivating crops, subsistence farm[ing], and all of those activities are directly impacted by changes in climate. If you were a woman and this is your life, you would notice when the water is getting harder and harder to find, when your crops are not doing so well, or just get withered by a drought, or your day-to-day life is likely going to be hit sooner than that of men.

In identifying several specific forms of labor, most often unpaid labor, that women perform across the world, Belina also indicates that many women might have a specific source of knowledge about climate change due to the tasks that they frequently perform.[6]

Existing research supports these assumptions about gender and labor. On average, women do a disproportionate share of unpaid care work in homes and communities across the globe. This care work consists of (1) specific household tasks such as cleaning, preparing food, collecting firewood and water; and (2) specific activities related to the wellbeing of children, the elderly, and the sick. Globally, men dominate the world of paid work and women dominate the world of unpaid work. In fact, women on average perform three times more unpaid work than men (UNDP 2015). Women frequently provide care within families because cultural norms and expectations shape the division of labor between men and women (Abramovitz 1996; Fraser 1987). This trend toward uneven care burdens holds across both wealthy and poor countries (Bittman et al. 2003; Coffey et al. 2020; Schaeffer

2019). This results in women in general, and mothers in particular, having some of the lowest rates of participation in the labor force, even in states in the global North (Daly 2000; Detraz and Peksen 2018; Marino, Romanelli, and Tasso 2013).

While there are considerable differences in how these trends intersect with political, social, and economic forces in society, gendered patterns of labor still result in specific tasks being more likely to be undertaken by women than by men. When these tasks involve resource use, there is one intersection between climate change impacts and gender, a women-as-resource-users storyline. Climate change results in resources such as water and fuelwood being scarcer in many places (IPCC 2014b). If it is primarily conceived as women's work to secure these resources, women will directly experience the negatives of climate change as they struggle to perform these tasks. Allison, a scholar in the United States, remarked that she thinks of women becoming "even more screwed than they are now, as climate change impacts make it harder to get clean water, or harder to grow crops, and that kind of thing." Gendered patterns of labor, therefore, are understood to contribute to distinct forms of climate vulnerability. This theme dominates much of the existing scholarship on gender and climate change (Dankelman 2002, 2010; Denton 2002).

The women-as-vulnerable discourse also highlights the physical harm that climate change can produce in women in a storyline that I term the gendered-physical-effects-of-climate-change. Interviewees explicitly reflected on the effects of climate change on gendered bodies. Some pointed out the physical impacts of phenomena such as resource conflict and migration for women. Shea drew on some of her experiences as an environmental lawyer in the United States working with women from Sudan to highlight links between droughts, physical violence, and migration. Along the same lines, Allison, a US-based environmental politics scholar, reflected on the ways that gender, migration, and exploitation intersect: "When you have populations, especially folks who are economic or political refugees, it's much easier, essentially, to get these people away from their support systems when their support systems are down, and to basically abuse them and exploit them when they don't have a lot of resources, when they really need the economic help or whatever. So, I think that migration is probably going to increase things like human trafficking. It may increase conflict, which will also be really bad for women. Particularly because things like rape are used so often in conflict these days." Allison draws a link between social vulnerability (i.e., lack of support networks) and physical vulnerability (i.e., risk of human trafficking or rape). This storyline says that women's bodies are likely to be increasingly at risk for violence because of the stresses that accompany climate change and because they constitute an already marginalized group—a claim that finds support in academic literature (First, First, and Houston 2017; Fisher 2010).

Violence and death during natural disasters vividly illustrate the gendered physicality of how humans experience climate change (Buckingham 2020; Young 2002). Indeed, multiple interviewees specifically mentioned women's vulnerability during disasters as part of their reflections on how gender might connect to experiences of climate change.[7] An example of this storyline comes from Ginnie, an environmental nonprofit worker from the UK: "I suppose in developing countries or in countries where women stay more at home, they're going to be more impacted by climate change. . . . They're the ones who are going to be most vulnerable because they might not have heard the news that those disasters are coming their way in the same way that men are more connected to their communities." Like Ginnie, other interviewees frequently mentioned both flooding and droughts in particular as disasters that intersect with gendered patterns of labor and vulnerability. In the case of drought, many participants mentioned the fact that women are typically the ones expected to provide water for the household, a task that becomes more difficult in times of water scarcity. In the case of flooding, on the other hand, the expectation is that women will be more in harm's way as waters rise because of gendered divisions of labor in homes and communities.

A large body of research has focused on gender and natural disasters as well (Arora-Jonsson 2011; Dankelman 2010; Demetriades and Esplen 2010; Detraz and Peksen 2017). In their widely cited article, Eric Neumayer and Thomas Plümper (2007) analyzed a sample of 141 countries between 1981 and 2002 and found that disasters adversely affect female life expectancy more than male life expectancy and that women have higher mortality rates in places where they are socioeconomically disadvantaged. This finding reinforces the idea that vulnerability is a condition made up of multiple forms of marginalization, including class and gender. One interviewee, Deb, drew on her academic work on natural disasters in evaluating the complexities of vulnerability and disasters.

> A lot of those that experience vulnerability, should you have a severe weather event, are about isolation. Now those who are isolated tend to be those less well embedded in communities, and if you have a single mother who is on housing benefit, is being housed in different places, moving around with lots of children—they are the most vulnerable because they are the least embedded and can't depend on family friends or others to respond. . . . In term of the effects . . . the simple explanation if you're a woman you have less money, less resources, less well-embedded, also the stigma of being poor, of admitting you need help, not willing to share problems with authority, it's income related and women still earn a lot less than men generally.

Deb's background in natural disaster research affords her unique knowledge of this form of climate change vulnerability. Her assessment of vulnerability contains several distinct storylines, including women-as-marginalized, women-as-caregivers, and women-as-poor. Her response, like Allison's mentioned above, also indicates that not all women experience vulnerability in the same ways. More specifically, women who tend to be at higher risk during disasters are the ones who are in a more precarious economic position, have children, and lack strong social support and community ties.

Natural disasters result in bodily harm during the actual event (i.e., drowning in a flood or dying during a tornado), but they also foster increased vulnerability in the social disruption following the event. For example, a few participants mentioned that women tend to suffer intimate partner violence (IPV) or other forms of physical harm during situations of social stress and that these increase as communities feel the effects of climate change. Marina, a US-based scholar, argued that disaster response tends to be male-dominated, hierarchical, and often blind to gendered challenges such as IPV—though this last point is changing. She felt that the characteristics of the response often reinforced women's vulnerability during and after disasters.

This gendered vulnerability to physical harm is illustrated by examples from a tsunami in South Asia in 2004 and Hurricane Katrina in the United States in 2005. Reports of IPV and other forms of abuse against women increased considerably immediately after the tsunami and hurricane (David and Enarson 2012; Fisher 2010). In the case of the tsunami, women and young girls who were separated from their families were particularly vulnerable to abuse. In a 2020 study of four Indian states affected by the tsunami, Smitha Rao found that rates of IPV increased in the ten years following the disaster with "social disadvantage" strongly predicting IPV risk post-disaster. In the case of Hurricane Katrina, a study of displaced women found that nearly one in five reported experiencing some form of abuse or violence (Anastario, Larrance, and Lawry 2008). There were also reports of increased levels of IPV following US Hurricanes Harvey and Michael in 2018 (McDonald 2018; Zurawski 2018). Studies have also found connections between IPV and other climate-related factors such as declining water availability (Whittenbury 2012). It is important to note that these patterns occur both in the global North and global South. Participants also provided examples of how women's marginalization facilitates unique forms of vulnerability in times of environmental change in multiple spaces. However, most of their reflections focused on women in developing countries, a point that will be discussed at length in later sections.

In sum, the women-as-vulnerable discourse is centered on the notion that various societies expect women to play particular roles in communities and in families. These gendered roles often include domestic labor and care work, which intersect with climate change impacts. All of these patterns combined put women at greater risk of climate change vulnerability. But what makes this story possible? If climate change exacerbates patterns of marginalization or exclusion, what are the social, economic, and political structures that make this so?

GENDER, CLIMATE CHANGE, AND EXISTING STRUCTURES

That climate change exacerbates existing patterns of marginalization or discrimination was a recurring theme throughout the interviews. It is also a common storyline in media, scholarly, and policy treatments of climate change impacts.[8] Several interviewees mentioned a connection between gender and climate change during such transformative processes as the Scientific

Revolution, the Industrial Revolution, and the Environmental Revolution—all dominated by men (Detraz 2017b; Plumwood 2002), which ushered in new patterns in the relationships between humans and ecosystems. This storyline was prevalent in response to my question of whether interviewees saw any connections between gender and the causes of climate change.[9] One repeated answer was that while it is inaccurate to claim that men "caused" climate change, they were the ones in positions to make decisions related to processes that contribute to climate change. Additionally, interviewees noted specific connections between masculinity and the economic, social, and political factors that have led to our current climate crisis.

Multiple participants focused on the male-dominated character of the Industrial Revolution as evidence that, at its most fundamental level, our current climate crisis has been driven by the decision-making of men. They argued that men dominate most of the high-carbon emitting industries, particularly the fossil fuel industry. Historically, they have made decisions about large-scale resource extraction and use, while it is marginalized communities that are now being disproportionately negatively impacted by the ramifications of those decisions. Several extended this storyline to reflections on the dominance of men in global capitalism more generally. Many were quick to point out that they do not necessarily think that women in positions of power would have made different decisions since they would have been responding to the same incentives as men. In this view, it is the nature of the position of power that facilitates the decisions to view the environment in utilitarian terms at the expense of sustainability. One interviewee, Glenda, an environmental lawyer in the UK, said,

> I mean the system we're living in is basically a system that's been set up for men. I'm not saying that women don't facilitate it and run along with it. But it's mostly, you know, the Industrial Revolution and the way we do things . . . and the idea that property is a very sort of masculine concept and, and contract law and this idea that there's always ownership involved.

Likewise Darcy, a US-based academic, expressed this view by saying,

> I'm drawn to thinking about the rise of neoliberal capitalism in the West, and global capitalism more broadly, and how it is that women are obviously

> increasingly engulfed in becoming exploiters, especially in the West and First World nations. But I think that historically they mostly found themselves on the side of those who have been exploited and so, I still think of that as, you know, [a] Western capitalist, patriarchal narrative more than anything.

Put differently, both Glenda and Darcy claim that while women are active participants in current economic structures, there are deep connections between these structures and masculinity that shape our positions within them.

The Scientific Revolution came up in the interviews in a similar way: participants argued that male-driven turning points in our history have fundamentally shaped humanity's trajectory, and that followed a specific exploitative logic about the environment and our place within it. Kylie, an environmental scholar working in the United States, argued:

> I think that it's probably rooted historically. Maybe back to the Scientific Revolution and how we "do" science in the West—and ideas about the material of the earth being insentient stuff that we can manipulate at will. . . . I mean, a lot, of course, can be said about Western science and how it developed from the Scientific Revolution. . . . And the "objective, modest observer who is unmoved by the consequences of scientific experiments" has shaped a lot of our technology since then and allowed us to create technology that externalizes negative consequences onto the surrounding environment.

Kylie thinks that the dominant scientific paradigm influences not only how science is conducted, but also how humans think about our obligations to nature and to each other. In her view, dominant approaches to science can contribute to environmental change because they see the environment as something open to manipulation. Environmental scholars have also pointed to how the Scientific Revolution continues to shape human-nature connections. Karen Litfin (2012, 420) explains that "after Descartes, Western science and philosophy generally understood the world as a machine, a view that has been increasingly globalized in recent decades." For instance, Francis Bacon famously claimed that "the world is made for man, not man for the world" (quoted in Hopwood, Mellor, and O'Brien 2005, 38–39).

While the mentions of the Industrial and Scientific Revolutions may not be very surprising in the context of discussions of structures that exacerbate inequality or marginalization, perhaps the presence of the Environmental Revolution on the list is unexpected. By the "Environmental Revolution," I am referring to the spread of environmental policymaking and advocacy around the world that accelerated in the second half of the twentieth century. While it is true that people have been calling attention to environmental issues for centuries,[10] over time societies began to see environmental issues as crucial social and political issues. It was not until the 1960s and 1970s that public demand for safer and cleaner spaces, coupled with the proliferation of environmentally focused nongovernmental organizations (NGOs), resulted in states paying increased attention to the environment as a political issue.[11] Environmental NGOs have played a central role in advocating on behalf of the environment and those communities that experience environmental change. The first environmental NGOs emerged in the late nineteenth century and included the International Union of Forestry Research Organizations in 1891 and the International Friends of Nature in 1895 (Betsill 2014). There are now a huge number of environment-oriented organizations that interact with the policymaking process in states across the world. The Environmental Revolution facilitated their professionalization and expansion as environmental issues gained traction (Wapner 2012).

While the number of environmental organizations has increased over time, most of the well-known organizations with cross-national reach trace their origins to middle- and upper-class white men in the global North. Gender norms played a role in the framing of the movements, as well as members' understanding of their place within them (Gottlieb 2005; Jones 2019; Nash 2001). In the late nineteenth century, when some mainstream environmental organizations emerged, concerns about masculinity were heightened as social and economic shifts associated with industrialization and urbanization resulted in new types of white-collar employment. These jobs were regarded as "soft" and incompatible with dominant standards of masculinity for the time (Mann 2011). Ecosystems provided realms within which to (re)connect with masculinity, as wealthy men of this era looked to wilderness adventures to foster "toughness" and "virility." While the types of environmental organizations and their foci have proliferated considerably since then (Doherty and

Doyle 2006), many well-known organizations have struggled to fully break from their traditional past.

This theme was discussed by several interviewees who work in the environmental nonprofit sector and who called attention the challenge that their organizations have faced in diversifying both their composition and their focus. The latter was described as involving things like moving away from a narrow view of the environment and toward a greater embrace of environmental justice concerns. For instance, Paige reflected on gender issues in her environmental organization in London as well as larger questions of gender "in the movement." She argued that

> the climate change movement is notoriously un-diverse. There are a decent number of women who work in climate change, but lots of charities wouldn't allow a woman to get to the senior leadership levels. So, I think there's an issue around women working in climate change. . . . I think just like any workplace, any move ups in seniority level, women tend to fall out. So, it is dominated by white, middle-class men. And that is deeply unhelpful for diversifying participation, and also it's a self-perpetuating cycle.

Even though many people like Paige who work "in the movement" recognize differential marginalization and the need to diversify, the male-dominated nature of the Environmental Revolution continues to mold the possibilities of climate change nonprofit work today.[12] This storyline in particular surfaced in several of the environmental nonprofit workers' descriptions of their experiences in their organizations.

Each of the three major historical developments mentioned in this section are often discussed as moving humanity towards "progress" in terms of objective scientific work, economic growth, and environmental action, respectively. What these depictions fail to tell us, however, are the consequences of these moves across various communities. Each one has intimately affected human relationships with nature, and each one is closely connected with masculinity. Gendered assumptions about domination and exploitation mean that it is not just that decisions were made by men, but that these decisions were made within a mindset that frequently privileged particular characteristics or processes. Examples include environmental exploitation or manipulation in the case of the Industrial and Scientific Revolutions, and hierarchical decision-making and a narrow idea of "the environment" in the case of the Environmental Revolution. We must reflect on who benefits from "progress" and who does not. How does marginalization manifest, and how does it intersect with large-scale environmental processes such as climate change?

CATEGORIES OF MARGINALIZATION

Social, cultural, political, economic, and physical factors influence who is most likely to be vulnerable to environmental change and how that vulnerability will manifest (Wisner et al. 2003). This section addresses some of these categories of marginalization as described by the interview participants. In most instances, they saw climate change impacts reinforcing or worsening existing patterns of inequality in societies. Those that are currently discriminated against, excluded, or susceptible to shocks in society will become more so in the face of climate change. Interviewees used storylines of race, class, age, disability, and place in order to describe these patterns of exclusion and marginalization.

Race

Racial differences in exposure to environmental ills is a consistent theme in environmental scholarship (Bullard 2005; Bullard and Wright 2012). Literature has also focused on the links between race and climate change vulnerability specifically (Shepherd and KC 2015). A few participants from the nonprofit sector discussed the steps that their organizations have been taking in order to reflect on race, exclusion, and climate change. For instance, Brenda explained how these debates have manifested in her US-based environmental nonprofit:

> We have worked very hard to center communities of color and populations that are going to be impacted first and worst by climate change and have done the least to cause the problem in the solutions. And by combining social justice and environmental justice along with the solutions to climate change, people who have been privileged and have had more than their fair share for a long time will necessarily get less. They will get what they should be getting, and that

> feels like . . . something is being taken away from them. And that is very difficult for them to accept. That's not just men, but it's people of privilege in general. But my experience is white men . . . have a really hard time accepting that their perspective is not the only perspective that matters, and not being the only voice that's being listened to.

Brenda reflected on the concept of race both as it influences who is most likely to feel the impacts of climate change, but also in the ways that addressing the differential burden of environmental change will require those in dominant positions to reconsider their status and make changes to distributions of power. Her mention of attempts to "center communities of color" and other affected populations calls attention to the steps necessary to achieve change in this area. Rather than privileged communities continuing to dominate climate change spaces as experts, policymakers, and changemakers, other populations need to drive decision-making forward.

Interviewees often listed women as one among a number of marginalized groups who are less responsible for the decisions that have caused climate change, yet more heavily impacted by it. Communities of color are another group within this category. Annise, a US-based nonprofit worker, said, "I think about women specifically, and how women all over the world will be inequitably affected. Much as how we speak about how communities of color are inequitably affected by climate change." Similarly, Shea drew on her experience as an environmental attorney as well as her work in the renewable energy industry in the United States:

> I'm very aware of how those most adversely impacted by the lack of sound environmental policies, and climate change reduction policies, are not only women but people of color—especially Latino people in this country. I'm very intimately aware that in Chicago the most polluted area is certainly Black and Latino. . . . In Atlanta, waste plants are always in the areas with poor people, who tend to be disproportionately Black and Latino. But working in mainstream nonprofits, especially on environmental issues, there is an unwillingness to address environmental issues and climate change through a race and class lens. Despite the fact that the ones who are most hurting are Black and Latino.

Shea further argued that women of color in particular face challenges achieving environmental justice. Her concerns about race and class, and the obstacles to fully engaging with these within "mainstream nonprofits" also connect with the previous discussion of the nature of the environmental movement.[13]

Kathryn, who works for a clean energy nonprofit in the United States, also brought up race:

> I've been doing some thinking, for instance, about the floods in Louisiana a few weeks ago—and thinking about race and about gender. So, thinking about how would Black residents potentially be more impacted than white. I don't have specific data on that but know . . . that . . . [there is] a combination of racial and discrimination issues that they face. You know, who has reason to trust the police. And who has family members with financial resources to support them. And, you know, a whole range of things. And everything that we know about housing and employment. And that may be similar, or parallel, or different—but related types of things to gender. So, . . . there are more single moms out there than there are single dads . . . and low-income families that are women-led. And that intersects with race, right? Because you've got this huge over-incarceration of Black men in our country. And so, impacts of poverty related to gender certainly affect people and how they respond to the impacts of climate change.

Kathryn's response brings up specific ways that marginalized communities have to navigate obstacles and inequities within existing economic, social, and political structures. This means that members of these communities have to deal with the impacts of environmental change while simultaneously addressing these obstacles. She explicitly mentions interactions between members of Black communities and police and the prison system as important considerations when discussing natural disasters in the United States. In these examples, gender, race, and environmental injustice are all inextricably linked. It is impossible to understand how particular communities will cope with climate change impacts without first understanding how they deal with other stresses, such as dire economic conditions and strained support networks. Although there has been some research on these issues in fields such as critical race studies, it has largely been absent from environmental debates within political science, and more importantly, environmental policymaking (Dillon and Sze 2016; Pellow 2016). Kathryn's comments underscore the necessity of an intersectional approach to climate vulnerability. It is not enough to think about gender, or race, or class on their own; rather, we must attend to how these categories overlap and at times reinforce each other.

Of note is that nearly all of the discussions of race in the interviews came from women currently working in the United States. There are likely several reasons for this, including the legacy of the anti–environmental racism movement that began in the United States. The term "environmental racism" was coined in 1982 by Benjamin Chavis, head of the United Church of Christ's Commission on Racial Justice, at a protest over the siting of a toxic landfill in a predominately Black, poor neighborhood in North Carolina. Chavis, a trained chemist, understood the lasting damage that a facility like the landfill could inflict on local residents. He defined "environmental racism" as "racial discrimination in the siting of toxic waste dumps and polluting industries, unequal enforcement of environmental laws, and the exclusion of people of color from environmental decision-making" (quoted in Schlosburg and Carruthers 2010, 13). This concept of environmental racism highlighted oppression, political disenfranchisement, and poor health. Research across several countries finds that race is one of the strongest variables for predicting where waste facilities or other polluting sites are located (Bullard and Wright 2012; Mitchell and Dorling 2003). Though concern about environmental racism was one of the earliest strands of the environmental justice debate, scholars and activists who use the concept today utilize a variety of perceptions about the nature of justice (Agyeman and Evans 2004). Hence, environmental justice is a concept that is central to understanding the numerous ways that the issues of environmental change, fairness, inequality, vulnerability, and marginalization are intertwined. At the same time, there is some criticism that gender has not been sufficiently incorporated into conceptualizations of environmental justice. Issues of race, class, and indigeneity have dominated most scholarly and policy discussions in this area, with gender and sexuality being frequently absent, although this is changing (Gaard 2011; Sze 2017).

Class

Like the other categories of marginalization discussed throughout this chapter, class influences vulnerability in important ways. Poverty connects to marginalization and exclusion in society and influences how people experience climate change (Hertel, Burke, and Lobell 2010; Leichenko 2014). Lack of economic resources undermines adaptation capacity and resilience to environmental change. Climate-related hazards affect economically marginalized communities both directly and indirectly. Climate change has repercussions for livelihoods, food availability and prices, and agricultural productivity. It can destroy homes that are built in environmentally precarious areas. These stresses worsen the marginalization that poor people already experience and can lead to chronic poverty for those in both rural and urban areas (Hardoy and Pandiella 2009; IPCC 2014a). While we might be tempted to think that this is a problem that is restricted to the global South, studies have found that this is the case in the aftermath of natural disasters in states within the global North as well. Junia Howell and James R. Elliott (2019) studied the aftermath of Hurricane Harvey in the United States and concluded that wealth inequity actually increases as government disaster recovery programs are administered in an area, again reinforcing the notion that vulnerability is multifaceted and forms of marginalization are reinforcing.[14]

Many interviewees outlined a specific link between poverty, gender, and vulnerability and saw poverty as one of the factors that exacerbated women's condition of vulnerability. For instance, Selma, a nonprofit worker in Germany, raised the issue of women increasingly being responsible for household budgets. As climate change puts increasing pressure on food prices and other necessities, many women will struggle to cope. She noted that this might occur as men migrate from rural areas for work. Paige reflected on the same issue in the context of the UK, where she works in the nonprofit sector, noting that as flood events become more frequent and severe, wealthy people will be able to move away from affected areas while poor people will not have that option. For her, this will likely worsen existing gendered inequality in the UK since women are "getting poorer" due to the government's economic policies. This point about poverty influencing who can relocate and who cannot was echoed in other interviews as well.

As noted at the beginning of the chapter, I noted how Farhana drew on her environmental organization's experience with disaster recovery projects in Bangladesh to argue that poverty influences adaptation to disasters as well as whether or not poor and marginalized women have their voices heard. She said, "I think the effects of climate change and the burden of poverty that comes as a result of climate change disasters is [*sic*] often more felt by women than perhaps men. That's not to say that it's exclusively felt by women, but we've found that in our disaster committees, for example, about 70 percent of those that are most affected are women." In her view, women's poverty also affects whether women are included in decision-making about climate change, as illiterate women will not typically be included in policy decisions.

Another storyline related to the issue of class concerns wealth, which most people strive for. This influences priorities and behavior that have helped usher in climate change and will need to be overcome in order to effectively address it. Some women argued that the desire for wealth is not just something to ensure one's survival, but rather a flaw in humanity that must be overcome. Along these lines, Kate pointed out that there is a sense of entitlement in the upper-middle-class US neighborhood where she works for an environmental nonprofit. In her view, wealthy people feel that they are already doing enough by recycling or donating to environmental causes, something she says she finds "ridiculous." For this storyline, wealthy people—and wealthy communities and wealthy states—are unwilling to confront their own position in processes that make others vulnerable.

Disability and Age

Vulnerability based on disability or age were two additional marginalization storylines that came up in the interviews, albeit relatively infrequently. Both the storylines reflect on connections between environmental vulnerability and the body in specific ways. Deb, a UK-based academic working on natural disasters, brought up both in talking about the way that vulnerability has been conceptualized in a rather limited way by the UK government. She explained that a person is considered to be vulnerable if "they are dependent on medical equipment, so somebody in a wheelchair, oxygen mask. If someone has a disability or someone who is older than sixty-five." While the "official" category of vulnerability has been oriented around the idea of health, Deb suggests that it is often much more about isolation or embeddedness—that is, whether or not people have access to networks that can provide help and support. This was echoed by Mary, who has worked for the UK government on various environmental outreach programs. She argued that the way governments assess vulnerability as well as value (i.e., property value) in order to determine who in a community is eligible for specific forms of aid is very narrow. Both Deb and Mary had backgrounds in natural disasters work or research that allowed them to point out specific policies or recent events. Both suggested that because of current policy approaches to vulnerability, rural residents might be hardest hit by disasters such as floods and might be least able to adapt. At the same time, they are often also ineligible for certain types of government assistance such as financial compensation. This highlights the importance of rethinking dominant environmental policies through different lenses that recognize vulnerability as a multifaceted, socially conditioned phenomenon.

Disability also came up in interviews in discussions of people's movement, either in city planning contexts or in natural disasters. For example, Mary told the story of an elderly blind woman whose house flooded. Neighbors banded together to "protect" the woman, even though she seemed to be getting along just fine. Mary specifically noted that it was largely the male neighbors who took it upon themselves to look after the woman. She likened it to a "male bonding thing." This relates to the idea of the "hero complex" discussed more in chapter 7. Mary was struck with the way that the community members perceived disability differently than the blind woman herself and regarded her as someone in need of assistance.

Disability is an often-overlooked category of marginalization that connects to climate change. There has been scant academic work on disability in disaster or climate change research.[15] Nor has it been prominently featured in more general, global climate change policy documents. Reports by the Intergovernmental Panel on Climate Change (IPCC), for instance, have devoted more attention to issues of class, gender, and place than disability. Occasional mentions of disability as one among a number of factors influencing vulnerability have little elaboration.[16]

The disability and vulnerability storyline focuses mainly on the perception or actual condition of those with disabilities being more vulnerable to natural disasters and the like. The age and vulnerability storyline features some of the same themes. Moreover, both children and older people are particularly susceptible to the impacts of environmental change (Watts et al. 2018). In the case of children, studies indicate that they bear a disproportionate burden of disease stemming from climate change impacts (Sheffield and Landrigan 2011). They also face unique challenges during events such as natural disasters, including restricted movement, less physical strength, and limited access to early-warning systems.[17] At the same time, they are typically excluded from decision-making about their future (Thew 2018). The distinctive nature of youth vulnerability to climate change impacts was the entrance point for thinking about gender and climate change connections for Mary, a UK-based researcher. She expressed frustration that academics seem to focus on gender or other categories when thinking about vulnerability, while often ignoring youth. Despite this perception of a lack of attention to youth, the issue of children's health and climate change did come up in a number of interviews.[18]

Finally, advanced age is often recognized as making individuals more susceptible to climate change impacts. This was specifically mentioned in a few interviews in the context of fuel poverty in the UK. While they were not necessarily directly tying the issue to climate change, interviewees reflected on how fuel poverty reflects gendered patterns of experiencing environmental issues in general. According to a research briefing for the UK Parliament, fuel poverty is measured in several different ways, but it generally refers to households that have to spend a large amount of their income on keeping their residence at a reasonable temperature. This is particularly a concern in Scotland, where in 2017 an estimated 24.9 percent of households were fuel poor (Sutherland et al. 2018). Yvette, who works for an energy nonprofit in Scotland, explained the connections between gender, age, and fuel poverty by noting that "you think of the elderly, you know women live longer, there tends to be an image of the old ladies in their freezing Highland cottages. And if it's the old and the young that are most vulnerable, and mothers are more likely to be home with kids as well." Here she indicates multiple dimensions of vulnerability, including age, gender, and motherhood status. This complex view of the sources and experiences of vulnerability was common among the interviews.

Place

The most frequently used storyline linking marginalization and vulnerability to climate change is place. Out of seventy-six interviewees, thirty-one mentioned at least once that women in "poor countries" or "developing countries" are be on the front lines of climate change effects. This storyline was sometimes linked to lack of education opportunities, lack of access to family planning, or exclusion from climate change decision-making. Talia, an environmental scientist working in the nonprofit sector in Germany, explained the situation in these terms:

> I mean, we're in Europe, we don't really realize that climate change is going on. We complain about the weather all day long, but this is not what climate change is about. And those who actually have to suffer from it are the people mainly living in the global South. And there, women are more likely to be responsible [for] taking care of the family and not having an office job or whatever. But it's more the down-to-earth realization that, "Ok if there's a drought coming up, I don't know what to do." Or "My cattle are dying and, I can't grow my crops." And essentially that means that food has to come from another source, but where from? You know? And these are all problems I think mostly women have to try and deal with because I think it's the poorest of the poor who suffer most.

Painting a specific picture of women in the global South, Talia depicted them as likely to live in rural areas, assume caregiving roles, and have direct connections to resource use. Additionally, this passage contains a few important themes that were brought up by multiple interviewees: (1) women's poverty in the global South, (2) gendered divisions of labor, and (3) differences between the global North and global South.

The global South in general was often conceptualized by the participants as lacking resources necessary for adaptation to climate change. States in the global South were frequently discussed as "developing" or "poorer." This is evidence that the previous discussion of class relates to economic power dynamics across the international system as well as domestically. At the same time, interviewees seemed to identify women in general as more likely to experience poverty, and women in the global South as struggling with this condition more than women in the global North. Ingrid, an environmental nonprofit worker in Germany, pointed to these connections:

> I think women are more affected by climate change and so they experience it stronger. I think it also has to do with education level, so in some developing countries, girls don't have the same chances to attend school as boys. They don't get to work in the same positions later on, and then they are typically the poorer people.

Here she links gendered access to education and employment with gendered patterns of poverty. In her view, this poverty results in women often having less adaptive capacity to address climate change. In general, most of the interviewees discussed how climate change effects would be felt by rural communities,[19] but this was particularly true when they brought up the global South. While they rarely said "rural women," they talked about tasks such as directly coping with drought or engaging in subsistence farming, which are rarely associated with urban life. There are important implications of this trope of the vulnerable southern woman which will be taken up in the following section.

As discussed previously, climate change impacts on natural resource availability and quality are frequently associated with women's vulnerability. In some cases, interviewees' perspectives were shaped by first-hand experience of fieldwork in other countries or by the stories of women whom they met in the course of their work in the global South.[20] Eva, a nonprofit worker based in Germany, reflected on her encounter with a woman from Malawi who took part in one of her organization's summer programs. She specifically focused on women's role in collecting fuelwood and how this becomes more difficult with climate change. Likewise, as noted above, Farhana acknowledged that her responses to my questions would draw on her work in Bangladesh with a UK-based nonprofit and on knowledge drawn from her Bangladeshi family. She mentioned women's tasks both in their day-to-day lives and during natural disasters as intersecting with experiences of climate change. Finally, a few interviewees noted the gendered divisions of labor they were familiar with in countries they spent time in doing fieldwork for academic research projects.

There was a much greater tendency for participants to talk about women outside of the global North than within it. Since all of the interviewees were currently working and living in a northern state, this is quite noteworthy. In fact, some said that they struggled to see how gender and climate change would be an issue for women in the global North. Marie noticed this tendency:

> It's interesting, though, how I tend to think of the global South first, then something that is close to me, but I think in our everyday life in the US and Europe, I'm not sure that the effects would be felt as strongly. I mean that there would be a gender difference on the effects of climate change because for us, it's more about, for instance, heat in the cities. I don't think that . . . heat would have different effects or would touch more women or men.

For Marie, climate vulnerability does not touch people in the global North as directly and not in ways that are as noticeably gendered as in the global South. However, there were some instances of participants first mentioning that women in the global South are the most vulnerable, then reflecting on their own environment or community. In the context of the UK, they mentioned flooding, rural areas and farmers struggling with weather unpredictability, and the potential for problems with sea level rise in low lying areas. In the context of the United States, they focused mostly on poor air quality or other forms of pollution in low-income neighborhoods and on different patterns of exposure to natural disasters. A few interviewees also mentioned that while those in the global North are not currently feeling the effects of climate change on a grand scale, they are next in line to feel what is already being felt in other places.

In sum, participants identified multiple categories of marginalization in addition to gender that intersect with climate change. Race, class, age, disability, and place were the most prominent categories. Interviewees typically reflected on mechanisms through which gender interacts with some of these other categories in order to make climate change experiences worse for those communities. The archetype of vulnerability in the global South is the poor resource user with family responsibilities. In the global North, it is the single mother. Heidi, a nonprofit worker in the United States, argued that "even in developed countries people who are the poorest and have the tightest resources, they are going to be the hardest hit. In this country, it would be single mothers that you often hear about." In both of these examples, the notion of marginalization has to do with the intersection of gender and class and involves gendered obligations of care work.

Because I have laid out the discussion of these forms of marginalization in discrete sections, it might appear as though the interviewees failed to make connections across them. This is not the case. For instance, Hildi, a nonprofit worker based in the UK, argued that it is essential to think broadly about the idea of marginalization and vulnerability:

> We talk a lot about who has to bear the brunt of climate change and within that again vulnerable groups are the ones that are hardest hit. And so there gender again I think plays part. But . . . I think it's still a wider social justice issue. So, while I would probably include certain groups of women, or minority groups such as transgender communities, in the more vulnerable sections, I think there's also a wider kind of global South/global North divide.

Her response identifies how people may find themselves in multiple marginalized groups simultaneously, as well as how global distribution of power might influence climate change vulnerability.

IMPLICATIONS OF THE WOMEN-AS-VULNERABLE DISCOURSE

An important question that guides my reflections on the implications of the women and climate change discourses is what is climate change? Women and climate change discourses shape understandings of women, but they also each serve as a representation or social understanding of climate change. According to Sherilyn MacGregor (2010, 229), "Climate change may be 'real' and have material manifestations, but it is also being shaped by social and cultural norms and discourses." Across the four women and climate change discourses, climate change is depicted as a global phenomenon that damages—it hurts bodies, harms livelihoods, limits futures. It is portrayed as an empirical global problem that needs to be understood and acted upon; climate change, like pandemics or terrorism, is a looming and consuming facet of humanity's present and future existence. It is represented as a space of work or action—people work "in climate change" as scholars and activists. This designation tells people which type of organization one works for or meetings one goes to or the type of research one does. Asking what is climate change permits us to probe climate change as a fluid and socially conditioned concept (Pettenger 2007; Weart 2011). The women-as-vulnerable discourse depicts climate change as an embodied phenomenon that is the product of human choices. It brings droughts and floods, makes water and food provision more difficult, and exacerbates burdens for those tasked with providing care for families. Interviewees regarded climate change as an existing source of hardship for those who are already marginalized, and something that will only intensify inequality and injustice in the future if humanity continues down our current path.

The plight of marginalized groups featured heavily in this understanding of climate change, with women at the front of the list. In the foundational book *Bananas, Beaches, and Bases: Making Feminist Sense of International Politics*, Cynthia Enloe (1990) asked what on the surface might appear to be a relatively straightforward question: "Where are the women" in international politics? Several decades later, it remains important to ask, "Where are the women" in the politics of climate change? Thus, we must consider both the spaces in which women are present and those from which they are absent. And we must think through their roles and actions as well as to question why these and not others? "Where are the women?" moves us to think critically about who the women are in the various discourse? Are they the ones who gender norms tell us to expect? What are the different categories of women's experiences that get depicted across the discourses? Participants in this study used the discourses not only to refer to themselves or other specific women but also to represent an imagined version of women.

So, where are the women in the women-as-vulnerable discourse? Women are in homes preparing food and tending to children and relatives, in fields and forests collecting resources, or in shelters dealing with the aftermath of disasters. They are absent from the boardrooms and policy tables where key decisions about fossil fuel and land use are made. They are likewise absent from the science labs or environmental organizations that help shape perceptions of "nature" or "environment" and humanity's relationships to these.

Interviewees I seemed highly aware of the links between marginalization and vulnerability. They reflected on ways that women tend to feel climate change "first and worst" because of their position in their societies. They referred to the ways that existing social, economic, and political structures limit women's agency and access to power or how poverty or racial discrimination inhibit adaptation capacity. Overall, women were not depicted simply as a group of victims, but rather as people struggling to cope with climate change. In reflecting on poverty in particular, interviewees tended to refer to the kinds of jobs that women tend to have (i.e., underpaid) or the fact that other obligations influence women's labor choices or options (i.e., caregiving roles).

Participants typically contextualized why certain women might be poor when others are not by considering their roles as resource users or single mothers. When considering who are the women, however, popular depictions of women in climate change discussions tend to paint them as rural, poor, excluded, and from the global South. Scholars and policymakers often discuss different roles or positions for women based on whether they are in the global North or global South (Arora-Jonsson 2011). Images used in government or intergovernmental organization reports on gender and climate change often feature women working in a field in a nameless African country or women in saris walking through floods. This seems to be the image that many interviewees had when I first asked them what comes to mind when they hear the phrase "gender and climate change": a rural, poor woman in the global South. These dominant portrayals of women jumped to mind, and it was only after asking some follow up questions that many offered specific reasons why women might be more vulnerable to climate change. It is also important to note that participants referred not to themselves using these storylines, but rather to women from "developing" or "poorer" states. While it is true that some interviewees were directly drawing on either fieldwork or projects they had done with specific rural communities as part of their job to answer my questions, the majority who used a women-as-vulnerable discourse were referring to a general idea of women from the global South."[21] According to scholars such as Chandra Talpade Mohanty (2003a), these kinds of depictions are both problematic and common. Representations of women outside of the global North typically lack context and assume a homogeneous, downtrodden other. This speaks to ways in which power flows though climate change spaces, a theme that will be discussed further in chapter 6.

The kinds of climate change spaces participants brought up, such as climate change negotiations, climate change panels at academic conferences, or government buildings where climate policy is debated, are located in the global North. There, dominant voices tend to be those of white men. They also have an oversized role in shaping discourses focused on women's victimhood, and these discourses become deeply embedded in shared understanding of climate change. People in the global South, particularly women along with other marginalized groups, are rarely in a position to shape discourses about experiences in the global North. But this is the fundamental thing about discourses—they offer shared understanding of phenomena such as environmental change, whether this understanding would be recognizable to the people cast within them. In this way, climate change debate and action typify unequal distributions of power around the world. This is in no way unique to climate change, but rather illustrates the continuity of patterns of domination in global political spaces and even who determines how scholars, policymakers, and the public understand the issues within them.

Intersectional analysis recognizes the need to reveal how varying forms of marginalization overlap and influence discrimination and vulnerability (Crenshaw 1994). The women-as-vulnerable discourse has enormous potential to overlook intersectionality. This would discount the ways that race, class, sexuality, dis/ability, and other factors interact with gender to shape the ways individuals go through the world. While there were some responses that lacked intersectional analysis, interviewees often avoided this tendency by reflecting on how gender and race or class build on each other to influence whether and how communities might be hard hit by climate change. It is interesting to note that participants located in different spaces focused on different aspects of intersectionality. While race was an issue touched on by some interviewees in the US context, class was a theme evoked by those working in the UK. Additionally, discussions of indigenous communities rarely came up, even though this is a major component of environmental justice conversations at the global level and the marginalized status of indigenous communities has significant bearing on patterns of vulnerability (Shearer 2012).

I originally decided to write this book because I wanted to understand how women see connections between gender and climate change. More specifically, I wondered whether women would use the vulnerability frames that I came across so often in news stories, policy documents, and even academic work. What I found was that participants did evoke the idea of vulnerability quite often, but that their understanding of vulnerability was more complex than I originally anticipated. In terms of frequency, the women-as-vulnerable discourse was heavily used across the interviews as well as the surveys that I conducted as a check on the interview responses. It was used 122 times across the interviews, and over half of the survey responses (eleven out of twenty) mentioned that women will be disproportionately burdened by climate change in their initial reaction to the phrase "gender and climate change," meaning that this is the first thing they thought of, before being potentially led elsewhere by any of my subsequent questions. In both the interviews and surveys, participants rarely couched women's positions in terms of victimhood (only four interviewees specifically used the term to refer to women feeling the impacts of climate change), but rather described the structures and conditions that make women more likely to be significantly impacted by climate change. In some cases, participants were using the women-as-vulnerable discourse to describe a hypothetical situation. In other instances, they drew on their research or projects they were involved in with their jobs. For instance, Marina and Deb referred to their research on natural disasters to explain how communities might experience environmental change.

Participants in this study were not unique to using a vulnerability discourse to describe women, particularly when reflecting on the consequences or experiences of climate change. In fact, high-profile global climate change texts also commonly use these kinds of discourses. Several IPCC reports along with the text of the Paris Agreement, for instance, stress women's vulnerability whenever gender is mentioned (IPCC 2014b, 2014c; UNFCCC 2015).[22] All of these texts frame women as facing social, economic, or political barriers in society, which shape their experience of environmental change and their ability to address it. The texts express a need for states to strive for gender equity in working toward climate change mitigation and adaptation, and these calls tend to focus explicitly on women's current exclusion or marginalization. They acknowledge women as either agents or potential agents, specifically in climate change adaptation. Yet, this agency is often depicted as being blocked by either direct or indirect vulnerability. One interesting aspect of these depictions is that although there is a great deal of anecdotal evidence about gendered climate change vulnerability, the international community lacks reliable gender-differentiated data on climate change experiences.[23] This was addressed in 2019 at the fourth session of the UN Environment Assembly in Nairobi with a resolution on gender equality in environmental governance that included a commitment to collect sex disaggregated data, along with data disaggregated by age and disability.[24] This kind of information is crucial for recognizing the specific ways that gender intersects with climate vulnerability. This recognition is the first step in addressing it effectively and justly.

So, does it really matter if we use vulnerability discourses to understand women and climate change? Discourses shape our very understanding of climate change as well as influencing policy debates in meaningful ways (Detraz 2017a; Stern 2006). Each of the four discourses discussed in this book offer both opportunities and obstacles to effective and just climate change approaches. On one hand, women in communities around the world *are* often among those who feel the impacts of climate change particularly acutely. Marginalization in the forms of less access to political and household decision-making, fewer adaptation resources, high potential for poverty, and gendered divisions of labor can contribute to women being on the front lines of experiencing the negatives of climate change (Alston 2011; Arora-Jonsson 2011; Dankelman 2010; Paavola 2006; Panitchpakdi 2008). Additionally, the stresses of climate change can exacerbate gender inequality in society. Recent empirical work points to a strong relationship between climate change impacts such as disasters and temperature rise and declines in women's economic and social rights. This is particularly the case in countries in the global South that rely heavily on agriculture in their economy (Eastin

2018). As climate change worsens and exacts greater burdens on societies, patterns of marginalization are likely to get worse.

Using a women-as-vulnerable discourse can therefore potentially highlight these gendered trends in vulnerability. As discussed throughout this chapter, vulnerability rarely results from one thing. Rather, it is the consequence of multiple social, economic, and political processes. Women and men tend to be differently placed in economic, political, and social processes in every country. For instance, women are underrepresented in politics in states around the globe. Women still tend to be overrepresented in low-wage or no-wage labor. Women tend to be underrepresented in many institutions that play large roles in processes of socialization, such as, entertainment or religion (Benería 2003; UNDP 2015). While it is true that the experiences of women in all societies differ, it is also the case that women frequently find barriers to accessing power in multiple spheres when compared to men in similar situations. Societal expectations about the appropriate or acceptable roles and responsibilities of men and women shape experiences (Detraz 2017b). The fact that women tend to face certain kinds of marginalization and discrimination in all societies means that climate change vulnerability is gendered. Justina Demetriades and Emily Esplen (2010, 133) stress that "where women and girls have less access to and control over resources (material, financial, and human), and have fewer capabilities than men, these impediments undermine their capacity to adapt to existing and predicted impacts of climate change, and to contribute important knowledge and insights to adaptation and mitigation decision-making processes." Large societal patterns of gendered marginalization and discrimination result in many women experiencing climate change vulnerability in ways that differ from men in the same social and economic position.

On the other hand, if women are simplistically portrayed in climate change debates primarily as vulnerable victims, then there is a danger that they become viewed as a population that needs saving rather than a diverse set of people who can potentially play roles in their own future. Feminist environmental scholars frequently express concern about the use of essentialized ideas of women as a vulnerable category of actors who lack agency

(Arora-Jonsson 2011; Cuomo 2011; Denton 2002; MacGregor 2009). Seema Arora-Jonsson (2011) argues that women are often simplistically portrayed as either vulnerable or virtuous in discussions about climate change. Portrayals of virtuousness are tied to the assumption that women are more environmentally conscious, especially in the global North. Depictions of vulnerability are typically linked to the poverty of women in the global South, along with their experiences of environmental threats like natural disasters. Arora-Jonsson claims that either of these portrayals can result in policymaking that raises women's responsibilities without corresponding rewards. Either women become environmentally friendly warriors who should take the lead in "fixing" the problem, or women become problems for other actors to help. Either way, there is limited space in current discussions for considering where women's potential environmental awareness, poverty, or vulnerability to disaster comes from. For instance, most climate change policy documents mention systemic discrimination and marginalization, but these appear as a giant barricade that people are politely trying to find ways around. There is rarely sustained engagement with questions of how to reduce the central forms of inequality that make some people more likely to suffer climate impacts than others. Climate change vulnerability is tied to the same social, political, and economic processes that make some groups of people more likely to be exposed to unhealthy environments (Bullard 2005),[25] more likely to die at the hands of their partners or family members during and after natural disasters (True 2012),[26] or more likely to be excluded from environmental decision-making (Agyeman, Bullard, and Evans 2003).[27] These examples all speak to the centrality of marginalization within the concept of vulnerability.

Another concern with vulnerability discourses is that they sometimes serve to reproduce larger patterns of marginalization within society (Thomas and Warner 2019). States and intergovernmental organizations often use conceptualizations and measures of vulnerability to determine adaptation policies. Some scholarship has documented ways that these adaptation processes can actually reinforce existing forms of vulnerability or reproduce them in other communities (Atteridge and Remling 2018; Barnett and O'Neill 2010; Warner and Kuzdas 2016).[28] The interviews with Deb and

Mary in which they critiqued the UK's conceptualization of vulnerability in the aftermath of disasters speaks to this trend of vulnerability storylines having an uncomfortable fit with policymaking. At best, policies leave vulnerable people out of adaptation initiatives; at worst, they contribute to making vulnerable people even more so. Likewise, Kimberley Thomas and Benjamin Warner (2019, 101928) highlight instances in which "powerful stakeholders use climate change to justify extensive coercion, and, in the most extreme cases, their adaptation efforts target social rather than environmental threats." They call this the "weaponization of vulnerability" and link it to climate security discourses. While most climate adaptation policies aim to limit or eliminate exposure to climate change effects and enhance capacity, some "novel forms of 'adaptation' identify people and social hazards (e.g., social disorder, violence, crime) as the principal dangers emanating from climate change" (Thomas and Warner 2019, 101928). This trend is particularly worrying considering that women's fertility is already closely linked to climate change by multiple states, IGOs, and NGOs (Sasser 2018). The southern resource user with care obligations and the northern single mother were two of the persistent images of vulnerability across the interviews. If a women-as-vulnerable discourse is automatically linked to a women-as-mothers storyline (discussed in the next chapter), then women's bodies can be at risk from the weaponization of vulnerability that has already occurred in policymaking around the world (Detraz 2021; Thomas and Warner 2019).

While I would argue that there are important negative implications to women in the global North frequently using the women-as-vulnerable discourse specifically to describe women in the global South, I think a few factors contribute to this portrayal. First, the majority of the women with whom I spoke did not have a strong background in feminist environmental work, gender studies, or any related field. For many of them, our discussion was one of the first times they had ever thought about gender-climate change connections. Given that most participants were giving spontaneous responses to my questions, they were likely drawing from the victimhood images that most of us are exposed to in the global North. When they reflected on these links, they used many of the vulnerability images and storylines already at hand (or in mind).

It also appears to me that some of what is happening with the vulnerability discourses, and a reason why so many interviewees talk about women in the global South, is that they are reflecting on vulnerability as a spectrum. This is why even as some of them talk about what climate vulnerability looks like in their own context, they still recognize that vulnerability looks different in different places. Some participants even acknowledged this. They reflected on the fact that their first inclination was to mention women in the global South and wonder why that might be. This is why pushing back against simplistic vulnerability discourses is so essential. Alternative discourses that consider women's position and contribution to addressing climate change and recognize the complex ways that gender connects to environmental change, policymaking, societal shifts, economic processes, and everything else that goes along with the super-wicked problem that is climate change are needed. It is essential to understand the wide range of roles women currently play and will play in the future and simultaneously to consider both the vulnerability that communities face and women's agency and expertise. The following chapters provide some alternatives to the women-as-vulnerable discourse, sometimes defiant and sometimes complementary.

4 WOMEN AS CAREGIVERS: CARE ROLES AND CLIMATE CHANGE

Gwen works for a US-based organization that uses a motherhood frame to discuss environmental concerns. Her organization considers climate change, air pollution, and other environmental issues as health challenges for families. She coordinates the organization's policy advocacy work, which is centered on "the big policy changes that we need to make" across the country. In her words,

> Our goal is to arm people with the knowledge about what the impacts of burning fossil fuels is on . . . our health—what's the impact on our children? What's the impact on our bodies as women? What's the impact on our children in utero? When they're young? On nursing moms? What are those health impacts that we are not talking about? That we're not driving that connection for people? And how do we educate them so that they want to take action?

Gwen mentioned that she was motivated to get involved in climate change work after reflecting on mercury levels in fish when she was pregnant with her second child. As she thought about the connections between pollution from coal-fired power plants and toxicity in fish, she became angry that women are told not to eat tuna when pregnant while companies are told not to pollute the air. This encouraged her to volunteer for the environmental organization where she now works.

When I asked for her reaction to the phrase "gender and climate change" she responded:

> For me, a lot of it is impacts of climate change on bearing children and having children. I'm an eternal optimist, but the choice that women have, or the choice

> women make about having children . . . we need to start thinking about what the impact is going to be on our kids. So, if I'm having a baby in nine months, that baby is going to be around for 75, 80, 90 years. What is the world going to look like? And as women, we now need to start thinking about that.

As seen here, the concepts of motherhood and care dominated Gwen's account of why she started environmental action, the activities of her current environmental organization, and how gender and climate change are connected. Each of these ideas represent storylines in a *women-as-caregivers* discourse.

The previous chapter described ways in which women are cast as vulnerable, as a segment of society whose marginalization and exclusion render them especially susceptible to the impacts of climate change. While many participants identified the single mother to be the epitome of the vulnerable woman in the global North, women's caregiving responsibilities were often identified as influencing how they are likely to experience climate change in communities around the globe. In this perspective, women are particularly vulnerable to climate change impacts because they perform the primary caregiving role for families. More generally, women are envisioned as playing care roles that intersect with climate change experience and action. These roles include women as mothers caring for young children, cooking for families, worrying about family finances and struggles, and tending to older relatives—each of which will be discussed below as connecting to climate change. The following sections outline the women-as-caregivers discourse as it was expressed in the interviews. It highlights storylines such as caregiving roles influencing women's thinking about climate change, motivating them to get involved in climate change action, and shaping their approach to parenting. It also details how actors use care/motherhood storylines strategically in climate change work. The chapter ends with a discussion of the implications of using a caregiving discourse to describe and understand women and climate change.

CARE AND THE ENVIRONMENT

The concept of care enters academic debates about environmental change in numerous ways, including the place of care in environmental ethics, assessments of climate change and care burdens, and the connections between caregiving and environmental attitudes and behavior. Various scholars have asked what role "care" has in environmental ethics (Donovan and Adams 1996; Friedman 1987; Plumwood 2006; Robinson 2011; Whyte and Cuomo 2016). This literature has evaluated how the idea of care shapes assumptions about both human communities and our relationships to ecosystems. Kyle Powys Whyte and Chris Cuomo (2016, 234) explain that "ethics of care understand moral agents as deeply and inextricably embedded in networks of ethically significant connections and conceive of caring as exercising responsibilities and virtues that maintain and positively influence relationships and general flourishing within those overlapping networks." A well-known example of a care ethic is environmental historian Carolyn Merchant's (1995, 217) partnership ethic grounded in the concept of relation between humans and the nonhuman world. This partnership ethic of earthcare has four principles: "equity between the human and nonhuman communities; moral consideration for humans and nonhuman nature; respect for cultural diversity and biodiversity; and inclusion of women, minorities, and nonhuman nature in the codes of ethical accountability." While she is careful to caution against essentializing the position of women to caregivers or mothers, she does focus on gendered experiences of care resulting in unique environmental knowledge and responsibilities.

This focus on gendered experiences and divisions of labor is common in much feminist environmental work, particularly early ecofeminist scholarship (Salleh 1997; Shiva 1989). For instance, Mary Mellor (2006, 102) argues that "women are not closer to nature because of some elemental physiological or spiritual affinity, but because of the social circumstances in which they find themselves." Accordingly, gender norms influence how humans experience times of environmental stress, particularly for those who are expected to care for others in their household or community. Fiona Robinson (2011, 155) argues that "soil erosion, deforestation, depletion of water reserves, climate change, and the increasing severity and intensity of 'natural' disasters—all of these processes have fundamental implications for our ability to care for particular others with whom we exist in a relationship." Environmental stress makes it more difficult to provide care as well as increasing the care burden for those expected to shoulder it, particularly women (Robinson 2011).

A large body of feminist work examines ways that climate change impacts increase the unpaid care burden for many women (Alston 2015; Arora-Jonsson 2011; Dankelman 2010; Detraz 2017b; Paavola 2006). And this is on top of already existing global discrepancies between women and men. A 2015 report from the United Nations Development Programme finds that women continue to have a large unpaid care burden: "Across most countries in all regions, women work more than men. Women are estimated to contribute 52 percent of global work, men 48 percent. But even if women carry more than half the burden, they are disadvantaged in both realms of work—paid as well as unpaid work—in patterns that reinforce each other" (UNDP 2015, 11). Scholars argue not only that the gendered care burden often left out of evaluations of current economic systems, but that it is also frequently missing in assessments of where humanity might go in the future. For instance, feminist economists have argued that most approaches to a green economy fail to consider the care crisis that exists alongside economic and environmental crises (Bauhardt 2014; Littig 2017).

An additional strand of literature reflects on the connections among caregiving, parenting, gender, and environmental attitudes (Blocker and Eckberg 1989; Blocker and Eckberg 1997; Davidson and Freudenburg 1996; Hamilton 1985a, 1985b; Levine 1982). Early studies by Lawrence Hamilton (1985a, 1985b) demonstrated that gender and parenthood interact when it comes to environmental concerns, with mothers being more concerned about environmental threats than either men or women without children. These results were echoed by other studies showing that there might be something unique in the way that mothers view or experience environmental issues (Blocker and Eckberg 1989; Davidson and Freudenburg 1996). The most consistent finding across the academic literature on gender and environmental attitudes is that gender is significantly associated with the extent of concern about specific environmental problems (McCright 2010; Mohai 1997; Xiao and McCright 2012), with women expressing greater levels of concern over "local" environmental issues with perceived health risks to family and community (Bord and O'Connor 1997; Mohai 1992).

Other work illustrates links between caregiving and the creation and spread of environmental norms (De Groot & Steg 2009; Matthies, Selge, and Klöckner 2012; Matthies and Wallis 2015). This work is premised on the idea that caregivers can influence the environmental attitudes and behavior of those they care for. For instance, in a 2012 study of Danish families, Alice Grønhøj and John Thøgersen found that adolescents' pro-environmental behavior is heavily influenced by norms within their family, specifically by how strongly it is manifested by their parents. Likewise, Gary Evans, Siegmar Otto, and Florian Kaiser (2018) find that people who grew up in the United States with mothers holding more pro-environmental attitudes tended to engage in more pro-environmental behavior as young adults.

Caregiver discourses are also a staple in media and policy depictions of the adverse effects of climate change. Stories of struggling mothers and vulnerable children are often repeated in global climate change debates. Portrayals of women carrying babies through floodwaters or walking hand-in-hand with a child through a parched field "reflect the durable binding of the lives and fates of women and of children in public imaginaries" (Rosen and Twamley 2018, 1). In fact, policy documents often point to an implicit connection between these two categories by lumping "women and children" into a single group. Highly visible sources such as the Rio Conference's *Agenda 21*, reports from the Intergovernmental Panel on Climate Change (IPCC), and the text of the Paris Agreement, among numerous others, repeatedly associate women with children as related, vulnerable groups (IPCC 2014b; UNCED 1992; UNFCC 2015).

However, motherhood discourses have been strongly critiqued by scholars who argue that they are often essentialist in their tendency to connect womanhood with motherhood (Bretherton 2003; MacGregor 2006; Sandilands 1999). For example, Catriona Sandilands (1999, xiii) explains that in "motherhood environmentalism," "women's concerns about nature, even if they have eventual public appearance and impact, boil down to an obvious manifestation of natural protective instincts towards home and family. It is all about threats to children and self-sacrifice for the sake of future generations." In this understanding of environmentalism, women are first and foremost motivated by their identity as caregivers and only secondarily as citizens, or nature enthusiasts, or justice seekers. Sherilyn MacGregor (2006) has critiqued these types of discourses (which she refers to as "eco-maternalism") as reinforcing a pattern that does little to destabilize existing gender codes or enhance democracy. These gendered assumptions place women into a category of benevolent mothers who deserve respect, but not as active and informed citizens who have much to contribute to sustainability debates.[1] Along the same lines, the tendency to speak of "women and children" reflects a problematic, uncritical supposition about the inherent link between the position of women and that of children. Feminist scholars such as Cynthia Enloe (1990) argue that conflating the position and agency of women and children can serve to reinforce patronizing assumptions about women and undermine their well-being in cases where an instinct to "protect them" disenfranchises or marginalizes them in decision-making processes.

There is a tendency within much environmental scholarship and activism to characterize humanity as heterosexual and cisgender, a trend critiqued in fields such as queer ecology (Gaard 1997; Seymour 2013; Sturgeon 2008). For instance, Cameron Butler (2017, 271) calls attention to the mainstream sustainability movement as failing to confront heteronormativity ("the positioning of heterosexuality as natural and normal"), cissexism ("the conflating of gender and sex and positioning of cisgender people as natural and normal"), and reprocentricity ("the positioning of reproduction or procreation as being central to all people's lives"). This strand of academic work problematizes or "queers" dominant conceptualizations of sustainability as well as care.[2] It challenges discourses that fail to acknowledge or address vulnerabilities faced by queer people around the globe. Additionally, it offers some alternative conceptualizations of care that reject a reprocentric criterion of earning care through reproductive potential and demonstrated usefulness (Seymour 2013).

PERCEPTIONS OF CARE

As seen in the scholarly literature, there are numerous ways the concept of care relates to environmental politics and action. Interviewees used several storylines related to the idea of care. First, humans might care about the environment, meaning that we are concerned about environmental damage. Additionally, humans might care for the environment by working toward environmental sustainability. Finally, we might care for other humans when we assume a responsibility to support, protect, and aid them in ways that become more difficult in the face of environmental change. It is this last category that dominates the women-as-caregivers discourse, though there are important connections across all three.

This discourse was used fairly frequently by participants. Of the seventy-six women with whom I spoke, fifty-two used a women-as-caregivers discourse at some point in their responses.[3] The majority of these specified the notion of women as mothers, while others reflected on care roles and climate change more generally. For instance, some used a women-provide-care-in-communities storyline to note the strong presence of women in activities such as community development and community-based environmental management. Elane, a nonprofit worker in the United States, stressed that this assumption about gender and community involvement has implications for development strategies. "Investors are more inclined to invest in women-based projects because they know that they will take the skills and they will nurture them and develop them and that they're more likely to stay in their communities. And men will be more likely to leave their communities, families, and take the skills and go elsewhere. So, there's this connection into strong development of the communities that is uniquely and typically gender-focused." Elane thus argued that there is an assumption that women are most likely to be the ones who stick around and work within communities. The corresponding assumption that men will be the ones most likely to migrate because of environmental change finds some support in the academic literature (Obokata, Veronis, and McLeman 2014; Wrathall 2012); however, this is not always the case. In general, it is difficult to assign a single "cause" for migration. The reality is often much more complicated (Afifi 2011).[4]

Other storylines focus on women's motivations for providing care and ways that they do so. Vicky, who works for an environmental organization in the UK, pondered the issue of how care expectations influence environmental action and even careers. She is an example of a participant who used a women-as-caregivers discourse quite frequently, yet never specifically referred to motherhood or children. She commented that while her career path brought her to working on a community climate change project, this has not been the case for everyone from her environmental science courses:

> I think women tend to be drawn more to working with people. Women are taught to care and just pay attention to individual situations. And I like working with people, and I see that as a way to make a difference rather than focusing on these, kind of, more abstract theoretical issues. But also, in my . . . work I'm surrounded by a lot of women who care about climate change and I've always found that interesting.

Her comment that "women are taught to care" indicates the socially constructed nature of care roles. She claims that while she has noticed large numbers of women in climate change spaces, but she attributes this more to the roles women are expected to play within society than to something inherent in them. These reflections illustrate a women-as-expected-to-care storyline.

In contrast, Constance, an academic in the United States, offered a very different explanation for why women might be motived to get into her field of atmospheric science. In her words, "It's always that idea that your maternal instincts come out. And this kind of work, you're kind of almost looking to take care of everybody and fix everything." Her argument that women are motived by "maternal instinct" assumed a trait inherent in all women, even without their necessarily being mothers. This notion, as well as Constance's description of women as being motivated "to take care of everybody," was echoed by other participants who used a women-as-nurturing storyline. For example, Sarah, a US-based academic, said, "I think I'm more nurturing. . . . I'm on more graduate student committees than anyone else."

It was more common for interviewees to describe women in general as nurturing. For instance, US-based academic Kristy said, "I feel like there are a lot of women who are interested in the environmental field, I mean, or at least environmental issues and sustainability issues, and so I don't know, maybe it's the more nurturing side." Lydia is a US-based nonprofit worker who emphasized community and familial bonds, said:

> I wouldn't necessarily call myself a feminist, but as a female I do see [women] being able to take a . . . different approach for climate change issues from a more

nurturing perspective. And for some reason I see females as being able to be the ones that basically can open their arms to more people, be more inclusive to people because women somehow foster bonds between people because that's who we are. We're mothers, we're sisters. . . . Forming bonds . . . is something that we do—something that we're good at.

Lydia tied women's "nurturing perspective" to their ability to foster connections with people—something she regarded as necessary for effective climate change action. Annise, another US nonprofit worker, also argued that women's nurturing approach is necessary for solutions. She specifically noted that "women have an innate nurturing characteristic, a more gentle nature, a more community-minded connectedness to other people. And I think solutions are going to require that."

Other participants rejected this stereotype. Ivy, an academic in Canada, observed:

> I have certainly noticed that the women that I talk to, very few of them tend to assert some kind of essentialist position, that, you know, women are inherently inclined to protect and nurture the environment, which I don't, you know, I don't agree with that position. And the majority of the women that I have interviewed . . . they do tend to have perhaps more of a critical perspective, they are much more critical of the chemicals that are being used, the pesticides, the herbicides, that sort of thing.

Ivy rejected essentialist representations of nurturing women while also indicating that she has not seen many women express this position in her work. She did notice gendered patterns in the agricultural communities she researches, but she links this to gender norms rather than biology. Likewise, US-based academic Kristy claimed, "I feel like it's very easy to slip into stereotypes, and I don't want to sound like I'm stereotyping people you know? I don't want to make it sound like all women are all, you know, touchy feely nurture people." What is noteworthy, though, is that both Ivy and Kristy acknowledged the widely held assumption that women are nurturing or that women play an oversized care role in society, even if they then reject or critique it. Many feminist environmental scholars have expressed frustration at the prevalence and persistence of these women-as-nurturers tropes

(Bretherton 2003; MacGregor 2006; Sandilands 1999), a topic discussed further below.

CLIMATE CHANGE AND PARENTING

For most interviewees, care roles in society revolve mostly around caring for children. There was very little reflection on caring for others, such as aging parents or other relatives, a fact that might bolster the claim of repro-centricity within environmental action and scholarship. Some participants mentioned that women think differently about climate change after having children. This storyline was used to reflect on their own experiences or to describe women in the abstract. The claim is that because women tend to have the dominant caregiving role in most societies, they think about environmental change through this lens. For instance, Linda and Jessie, both advocates for children's health, reflected on how their role as mothers shapes their climate change work. Linda argued that having kids made her want to work harder on these issues. Jessie said that "as a mother it's a moral obligation to work for better health." She argued that she cared about the environment before she was a mother, but she thinks about it differently now. When she became a mother, she committed to protecting her child from dangers such as falling off chairs or other forms of physical harm. She viewed climate change as an overarching danger that she tries to protect her child from.

A related storyline is how climate change influences parenting. According to Sarah, an oceanographer working in the United States, "Being a woman and a mother makes you think differently." She noted that having her daughter "made me think more broadly" about the issue of climate change. She told a story about how depressing conversations about climate change between her and her husband (a climate scientist) have influenced their daughter. She referred to her child as traumatized by these discussions. Her daughter has told her that she does not want to have children of her own. This experience made Sarah realize that how caregivers talk about climate change to children matters.

Participants also tied climate change to parenting (and specifically mothering) through the idea of having a stake in the next generation. According to Ashley, a UK-based environmental ethics scholar:

> I started thinking about parenting and climate change when I had children. You sort of start reflecting on things that relate to yourself in some ways. I think yeah . . . working on climate change and having a stake in the next generation, the combination can be quite stressful.

Similarly, in response to the question "What comes to mind when you hear the phrase 'gender and climate change'?" Kylie, an ecology and religion professor in the United States, had the following to say

> I think there can also be the concern of a mother, about what the future will be of her children, and certainly concern for children is not necessarily gendered, but often times women have a larger role in caring for and educating their children. And I think giving birth to a child creates a particular connection to the future, a visceral symbiotic action that I was completely unaware of until I went through that experience. And so, I think that could be another aspect of gender and climate change or women and climate change. Just really having the very visceral investment in the future well-being of children.

For Kylie, the physical act of giving birth made her uniquely aware of the connection between this generation and the next. While others were less explicit about the physical nature of motherhood, they too reflected on the long-term implications of climate change impacts through their own lens of being a caregiver.

WOMEN'S CAREGIVING AND CLIMATE ACTION

Another theme that emerged from the interviews is that motherhood or the role of caregiving might make women more conscious of climate change and more likely to get involved in climate change mitigation or adaptation work. This is a women-as-aware-and-involved storyline. Aida, an environmental engineer based in the United States, said that "I guess in the 'women are nurturing' type way then they may care more to do what they can . . . or do their part [to address climate change]." Like some interviewees discussed above, Aida attributes women's motivation to engage in climate change action to their being "nurturing." Additionally, Elane, who works in the environmental nonprofit sector in the United States, argued that in her experience, mothers are more likely to get involved because they have first-hand knowledge of possible adverse consequences of climate change for their children:

> I think also that you have women who are, when you talk about pollution, I think . . . women tend to be more vocal advocates on behalf of their children. I think that some of the moms inspiring grass-root organizations that have been incredibly effective are frequent mothers-out-front who have wanted to ensure that their children have clean air or clean water . . . who are the ones most likely to take their kids to the pediatricians for asthma attacks . . . who know intimately or are more likely to stay home with their children when they're sick. So, there is this particular interest in focus that I think when you inform and educate women about these issues, you have a tremendous group of advocates on your side to really champion real solutions.

Elane uses the term "advocates" to describe women as mothers and mothers as agents for climate change action.

This sentiment was also brought up by Haley, who works for a US solar energy nonprofit and reflected on her own position as an expectant mother:

> I think, and part of this is me speaking as somebody that's an expectant mother, the older I get . . . the more real it becomes for me. Just from the simple fact of, you know, it's our future and it's our kids' future. And I know from speaking with other women, a lot of them feel the same way. It's a stereotype that women are more empathetic or sympathetic towards these sort of causes, but I think there is some truth to it. When we look at, "okay well, we have families and we're bringing kids into the world." And that's not to say that dads don't feel the same way, but when you're looking at who is actually motivated to spend their time volunteering or working on a campaign that's going to affect climate change, everybody has a motivation, and I think moms are people that are connected to their kids, which tends to be a lot of females.

According to this perspective, women are potentially more likely to get involved in climate change causes because of their role as mothers. While Haley, along with other participants, acknowledged that fathers also care about their kids' future, she indicated that there is something unique about what motivates mothers. Ginnie mentioned that of the sixteen thousand people who have taken part in her environmental organization's affordable energy programs in the UK, around ten thousand were women. She reasoned, "I guess that's because women are more likely to ask for help, or women are more likely to attend groups. . . . So, a big group of people who are vulnerable to poverty are people who have got young children and single moms. So, we do a lot of family work, in terms of women who come to those groups." This association between climate change and parenting certainly played into the path of Kate in her work with an environmental nonprofit in the United States. She explained:

> I was sitting at home and I was feeling really nervous about climate change in general, and I felt very isolated because I have two small children. And so, I started looking online for . . . I might have Googled like "good news climate" or something like that. And I feel like I was a little nervous and scared and I fumbled upon the page for our organization and I just started volunteering. I think they pegged me as the joiner that I am and I feel like it was an event where I showed up to volunteer, they were like "Hey, do you like this? Do you want to do this?" So that's kind of how I landed in this world.

She went on to say that thinking about the enormity of climate change made her feel "hopeless" and that she hated feeling that way, "especially as a mom." That is what ultimately drove her to get involved in her environmental organization. Sharon, who works in the US environmental nonprofit sector, brought up this trend of women using motherhood discourses to identify their own sustainability actions. "Not to say that there aren't some great men out there who are full believers," she said. "But, it's just to me [. . . I see a lot more women using that messaging in their own explanation of why they care about the issue."

Paige, a nonprofit worker from the UK, was one of the interviewees who pushed back against the assumption that "caregiver" automatically means "mother":

> One of the things I think I get cautious about is the assumption that women care more because women are caregivers and mothers. I always think that's a little bit over-simplistic. I think fathers care just as much about the children as women do. Although I think you do tend to see possibly slightly more, both in our work, for women having more collective concern. We often see a lot of

women coming forward for community-based projects. They have a heightened sense of community responsibility. . . . They seem to see it as more their responsibility to make sure everyone is ok.

Although Paige recognized that men also care about their children, she reinforced the sentiments expressed above that (1) it is still women who are more likely to take action on climate change by working on nonprofit campaigns and the like, and (2) women still assume (or are expected to assume) care roles for the community. Meg, a nonprofit worker in the United States, argued that this has been her experience in her work on climate change issues. In her view, women are motivated to try to save the planet for their family and friends, while men are motivated to save it for "the glory." While participants who used the women-provide-care storyline differed in whether they saw this as because of society's expectations or because of something inherent in women, they all argued that there are gendered patterns in who shows up for climate action and why.[5]

Caregiving and Health

Participants noted that gendered expectations result in women having the primary responsibility for addressing the health consequences of climate change for families. Effects of climate change, such as a lack of safe drinking water and food, poor sanitation, more frequent extreme weather events, a lack of shelter, changes in exposure to toxic chemicals, and population migration (Rylander, Odland, and Sandanger 2013; Sheffield and Landrigan 2011; Watts 2015), lead to a range of adverse health consequences, including changing disease patterns and morbidity. Women (especially pregnant women), babies, and children are among the groups that are particularly vulnerable to several negative health effects of climate change (Watts 2015).[6] Recent work estimates that climate change impacts were responsible for more than 150,000 deaths worldwide in 2000. Of these climate-related deaths, 88 percent were children (Sheffield and Landrigan 2011). Scholars identify links between natural disasters and malnutrition as well as disease in pregnant women, infants, and children (Rylander, Odland, and Sandanger 2013). Other research focuses on issues such as water salinity and maternal health (Khan et al. 2011), higher temperatures and low birth weight (Deschênes,

Greenstone, and Guryan 2009; Ngo and Horton 2016), and agricultural changes and low birth weight (Bakhtsiyarava, Grace, and Nawrotzki 2018).

A few interviewees reflected on these connections by discussing health, air quality, and climate change. Asthma and other allergic airway diseases are a serious threat to global public health. The rates of these conditions have increased dramatically since the late 1990s and are predicted to worsen with the advancement of climate change (D'Amato et al. 2015; Sheffield and Landrigan 2011). It is estimated that over 300 million people worldwide are currently affected. According to the World Allergy Organization, greenhouse gases in the atmosphere have already warmed the planet significantly, "causing more severe and prolonged heat waves, variability in temperature, increased air pollution, forest fires, droughts, and floods—all of which can put the respiratory health of the public at risk. These changes in climate and air quality have a measurable impact not only on the morbidity but also the mortality of patients with asthma and other respiratory diseases" (D'Amato et al. 2015, 1). One interview participant, Jessie, said that she has already witnessed the impact of poor air quality on women's time commitments through her work with an environmental nonprofit. She argued that when kids have asthma, mothers are typically the ones to care for them. While she said that she is happy to perform this role in her family, she also recognized that it could have negative consequences on her job and other commitments. "When my daughter is sick, it's me who stays home. . . . Women, we bear the brunt of it." Her response focuses on women's double roles as mothers and workers. It reveals a tension between the gendered divisions of labor in many families and women's paid labor, as women are still primarily expected to complete a great deal of unpaid labor (UNDP 2015).

Jasmine, a US environmental scientist in the US, and reflected on gendered experiences of climate change in terms of climate change–related health problems for pregnant women.

> I think also reproductively. [The] Zika virus outbreak shed some light on that. I think with climate change we have these disease vectors that expand and spread and . . . women have to deal with those consequences if they're infected with, like, the Zika virus. . . . [We] bear the brunt of taking care of children. And also . . . you see a lot of policies like "Oh, women have to refrain from having sex" or

> like "refrain from having children" because of this disease that's possibly being spread due to climate change. But no policies are really targeted towards men.

These comments call attention to the ways that climate change impacts pregnant women's bodies, but also the expectation that women's bodies are an acceptable site for policy attention, while men's are not (Detraz 2017b; Hartmann 2010).

Climate Change, Vulnerability, and Women's Labor

Interviewees often used the women-as-caregivers and women-as-vulnerable discourses in tandem to stress that women's unique vulnerability to climate change intersects with their expected care roles in a women's-environmental-and-care-labor storyline. They pointed out that many women, particularly those in economically precarious positions, feel the impacts of climate change through reduced access to water, energy, and food, which in turn influences their education, livelihood, and health. These factors compound existing inequalities within their societies, including reinforcing gendered roles such as unequal divisions of household and community labor (Alston 2011; Arora-Jonsson 2011; Dankelman 2010). Interviewees noted that climate change that adversely impacts water and food availability (access and cost) and the health of family members will likely result in additional unpaid work for women. Farhana, an environmental nonprofit worker in the UK, reflected on this in the context of her recent visit to Bangladesh for her organization. She recounted the struggle of women to grow crops and food as well as provide safe housing for their families in the aftermath of Cyclone Aila. Her responses frequently invoked the idea of women, and specifically mothers, working to overcome the challenges trigged by natural disasters.

One interesting facet of discussions of women's labor that emerged in the interviews is the idea of emotional labor that often falls on women's shoulders, a women's-emotional-burden storyline. Brenda mentioned this in the context of worrying about the impacts of climate change in general:

> I think there is a component of mental health that we don't talk about very much related to climate change and I think that women probably bear a lot of the brunt of that because they take care of a kind of emotional . . . they do the emotional labor of maintaining a household and maintaining children and thinking about the future of children and more probably than men do,

> on average. And I think that experience of climate change or looming climate change, because I think, probably, in the moment you're dealing with whatever crisis is in front of you and everybody's dealing with it in the same way, but if you're focusing on what may happen in the future or what will happen in the future, that is probably more on the shoulders of women than of men.

Likewise, Sina, a nonprofit worker in the United States, sketched these connections through a discussion of food security and psychological burden of women. She claimed that since women are "the backbone of the family" and tasked primarily with taking care of children, they have to worry about possible food shortages spurred by climate change. "I think in terms of psychology; it is affecting the women more as they experience that." Sina repeated this storyline tying women's care duties and increased vulnerability to a psychological toll of climate change: Usually, the women are very concerned about the future of their children—so, like, "This is what is going to happen, and I don't want my children to face that kind of condition. So, what should I do to change that, or to minimize that? . . . They're the one who is psychologically impacted." In addition to noting this psychological or emotional burden on caregivers, she also argued that "because they're very concerned and care for their family, they're the ones who . . . actively offer themselves to either to learn or to do something." Like several other participants, she said that this care role leads women to become more involved in trying to address environmental problems such as climate change.

Emotional labor even came up in the context of who performs what tasks in an office full of climate change activists. Swati argued that women are the ones "who tend to play the caretakers, like making sure there are birthday cards for people, making sure the kitchen is clean, making sure that if someone is leaving there is a going away party, things like that—that's not in any of our job descriptions, and the men kind of . . . it just happens, so why would they ever have to step up?" These reflections on gendered divisions of labor in society and families[7] were one of the first things most participants mentioned in response to the question "What do you think of when you hear the phrase 'gender and climate change'?"

Some participants reflected on the negative implications of gender stereotypes for both women and men who are taking action on climate change.

They indicated that even though thinking about climate change takes a serious emotional toll, men may not feel as comfortable talking about or expressing this aspect of their work. Participants such as Swati and Janice said that women in their office will cry tears of both joy and sadness but the men in their office do not cry in front of the group. Janice observed:

> I think that the women will talk more about the loss, actually verbalize this experience of loss, but I do not in any way think that that means that the men are not experiencing it at all. And in fact, given that men have sort of overarching sense of responsibility, I actually think that the men, for men, their pain may be more extreme even than ours. Because we talk about it, and we cry. Whereas I see them just look stricken and helpless, and this is the dynamic of working in climate is you can't be hopeless, you can't be hopeless.

For Janice, gender stereotypes in climate change work result in men being unable to express the significant emotional toll of climate struggles. Scholars have documented similar trends in the aftermath of disasters, when men are often depicted by the media as stoic and resourceful and women are portrayed as tearful victims (Emmanuel and Enarson 2012; Enarson and Morrow 1998). They argue that this portrayal is damaging, as men are denied the space or ability to be emotional in the aftermath of natural hazards. Scholars have also reflected on this trend within environmentalism in general, with certain forms of masculinity necessitating a sense or at least a projection of control (Norgaard 2011). This version of stoic masculinity may also relate to some participants feeling unable to express emotion in the spaces of science. In such a traditionally male-dominated sphere that prizes objectivity, displays of emotion are understood to be taboo—at least they have been in the past. In her work on gender and water, Farhana Sultana (2011, 164) says that emotions are an essential component of environmental issues. Scholars can assess "the multiplicity of ways that emotions are not feminine, but are constituted as a result of spaces, places, bodies, and experiences." What is problematic is that many societies in the global North have come to associate emotion—a human experience—with femininity. They have also come to regard the display of emotion as something unprofessional or uncomfortable. Grief, loss, and fear of climate change are obviously a very real aspect of climate change work for the majority of these participants. Likewise, there is an ever-expanding academic literature on climate change's impacts on mental health (Berry, Bowen, and Kjellstrom 2010; Bourque and Willox 2014; Cunsolo and Ellis 2018; Obradovich et al. 2018; Searle and Gow 2010). It is essential that people be given the tools to acknowledge and process the feelings that accompany climate change while also working to address the gendered material and structural challenges presented by climate change. This is particularly important for those who spend a great deal of time thinking about the topic.

WOMEN AND POPULATION GROWTH

Another important thing to note is how the women-as-caregivers discourse gets connected to the drivers of climate change in population discussions, a women-as-child-bearers storyline. This has been a high-profile and controversial topic since the 1960s (Cripps 2015; Detraz 2017a; Foster 2017; Sasser 2018) and going back centuries through Malthusian concerns about population growth, food insecurity, and environmental change. Women's fertility gets linked to the causes of climate change, but also to human vulnerability and lack of women's empowerment. In policy documents such as reports from the IPCC, reducing fertility is presented as a "health co-benefit of climate change mitigation measures" (IPCC 2014c, 742). That is, it is a win-win: women's health is improved at the same time that greenhouse gas emissions are reduced. This is consistent with the treatment of population growth discussions in *Agenda 21* and other high-profile environmental sources (Foster 2017; Sasser 2018).

The topic of population growth came up a few times in the interviews and closely echoed the themes above. In one instance, it was brought up by Ashley, an environmental ethics scholar, who was working on a project on population and climate justice. She asserted that when you examine the "link between population growth and climate change, then you start thinking about women's empowerment, or lack thereof, and how that's a key factor in both." Ginnie, who works for an environmental organization in the UK, similarly cast this discussion in terms of women's empowerment:

> If women were educated or had the contraceptives available, then they would have control over their family planning. And so unplanned pregnancies would, you wouldn't have as many of them, so that gives control back to the women which might be a benefit for their life in terms of being able to get an education or moving further into their career. And in terms of climate change impacts, having a child is one of the most impactful things you can do so if we can sort of have win-win scenarios where you're educating women, having controlled families, then maybe population growth might be checked and then resources might be better managed.

Both of these participants saw women's empowerment as an important way both to mitigate climate change because of their status as child bearers and to improve their position in society. Likewise, many in the international development sector use a similar population, environmental health, and empowerment frame. According to population scholars like Jade Sasser (2018, 2) this approach typically contains an implicit assumption that actors in the global North will be involved in providing poor women around the world with universal, voluntary access to contraceptives and education opportunities, which will enable them to "make decisions about their childbearing in ways that affirm their human rights while benefiting the environment by decreasing human numbers. In this schema, fewer people will consume resources and use polluting technologies, relieving pressure on the earth and its atmosphere." However, many scholars critique this emphasis on altering women's fertility as a means to some other end (Hartmann 2010; Ojeda et al. 2020). It is different from a specific, unique concern about women's reproductive health. There are plenty of discourses oriented around a particular focus on ways to foster greater social and health outcomes for women around the world. These discourses center the wants and needs of women as the goal in and of itself rather than linking it to global concerns about environmental sustainability, overpopulation, or resource use. They respond to the very real need for maternal, infant, and childhood health services as well as poverty reduction and family planning services in countries in both the global South and the global North.

In contrast to focusing on women's empowerment, Kate, a children's health advocate in the United States, rejected elements of the debate that she viewed as blaming women's fertility for environmental damage:

> I feel like sometimes . . . the finger gets pointed toward people who have decided to have children, and how that possibly is contributing to climate change and the use of resources and things like that. . . . I don't think that is fair in a way, because I feel like in society men and women have become the ultimate consumers. Both genders consume at rapid rates whether there are children involved or not, and I don't think that one consumes more than the other.

For Kate, consumption is a more central concern than population, and both men and women engage in consumption. While the storyline of women-as-consumers will be taken up in chapter 6, it is important to point out that here it was used in ways specifically connected to discussions of population growth. Another example comes from Nicola, a US-based environmental nonprofit worker, in talking about changes in her thinking about having children. She said that she did not intend to have children for a long time because it seemed like a responsible choice "to not bring another consumer into the world." Over time, she began to consider whether having a child meant that she "could bring a positive change into the world" by raising someone "who could continue the work that is needed to, sort of, transform our society away from a capitalist consumer society into one based on . . . justice and sustainability." She said that she began to think about "how it isn't black and white in terms of having kids being just another, sort of, suck on our resources or toll on our resources. But a potential solution to the challenges that we're facing." This transformation in her thinking hinged on her reflecting on the power of parents to guide their children, rather than all humans just being consumers.

These personal decisions about whether to have children were mentioned by other interviewees as well. Brittany, a nonprofit worker in the UK, posited that since women are the ones who birth children, they are the ones who most immediately deal with questions about whether to have children or how many children to have. She also pointed out that population is something that is not discussed a great deal, at least in her experience.

> It's quite uncomfortable. Quite an uncomfortable thing to talk about, what would life with your children be in the future? And whether it's then right for you to have children anyway because of that, and because of that contribution

> to future society. I think it's an uncomfortable topic for people to talk about—it's not, like, free and easy.

Sharon said that she has heard uneasy conversations in her US-based environmental organization about whether women, including herself, were planning to have children. In addition to the strangeness of discussing extremely personal choices with colleagues or acquaintances, such a conversation takes on a totally different character among environmentalists since population has been so firmly linked to environmental debates since the 1960s and earlier.

USING MOTHERHOOD/CARE DISCOURSES STRATEGICALLY

While caregiver discourses are often used to indicate a specific form of vulnerability for women, in some instances they are invoked strategically to get people to care about the future marked by climate change. For instance, Mary Robinson and comedian Maeve Higgins launched a climate change–themed podcast in 2018 called "Mothers of Invention." Robinson, the former president of Ireland, is well known in the international community for work on social justice and environmental issues. In introducing the all-female series, Robinson explained that "climate change is a manmade problem that requires a feminist solution. . . . What we are hoping to do is create a movement. Climate change is not gender-neutral—it affects women far more. So, this is not about climate change, it is about climate justice" (quoted in Harvey 2018). The title of the podcast is a play on the often-used phrase, but it also relates to existing social expectations about motherhood being a common feature of women's lived experiences. Additionally, numerous environmental organizations use the theme of caregiving or parenthood. While some climate change and general environmental organizations tap into parenting discourses more generally (e.g., Climate Parents, Parents Roar, Canadian Parents for Climate Action, Grands-Parents pour le Climat), others are specifically mom-themed. These include Moms Clean Air Force, Cool Mom, Mothers Out Front, The Mother's Project, and Climate Mama. These organizations boast a million members collectively. Kelsey Wirth, the founder of Mothers Out Front, has said in interviews that she regards the organization of mothers into an activated constituency to be no less than reinvigorating to our democracy (Wroth 2016).

This trend of strategically using motherhood or care narratives has long been implemented by women activists. In countries such as the United States, middle- and upper-class women have often participated in environmental initiatives and evoked the image of motherly duties. During the Progressive Era at the turn of the twentieth century, women's clubs used the frames of motherhood or household needs to put forth the claim that they were naturally inclined with the experience and temperament necessary to clean up their communities and the country. These frames also fit societal expectations about the appropriate role for women and thereby allowed them to engage in activism without presenting a radical challenge to gender norms.

Although the clubs were segregated by race, both white and Black women's clubs drew on motherhood narratives in their activism, albeit in slightly different fashions. Black clubwomen "drew on the legacy of Black women's roles as 'othermothers' and caretakers of the entire community" (Stover and Cable 2017, 688). Harkening back to times of slavery and decades after when Black families were often separated and Black women cared for the children of other women, clubwomen presented themselves as caretakers of rapidly expanding urban communities (Stover and Cable 2017). These types of socially salient narratives continued to accompany women's participation in the proliferation of environmental activism in the 1960s and 1970s, particularly in environmental justice–focused organizations.

My interviewees argued that people are much more receptive to messages about climate change when it is discussed in terms they can relate to. This response came up repeatedly in interviews with those who work in environmental nonprofits in particular. Brenda, for instance, explained that she sometimes uses images of people's past and their children's future to get them to connect with the idea of climate change in her work in Seattle, Washington:

> We want our children to see the Washington that we find so special. We want our children to be able to go up to the mountains and see snow in the wintertime. We want our children to go, to be able to go out to the beaches and see the sea stars and see the, you know, gather oysters, or whatever. You know, all of the things that climate change is impacting. . . . So, those are the arguments

> that we have found are the most resonant with people who don't necessarily automatically . . . who aren't already believers in climate change.

Meg, another nonprofit worker in the United States, also used the argument that "I care about preserving this earth, my grandchildren need it. I don't have children, I don't know if I ever will, but I think it's something that I use because it's an easy connection." Meg's is a clear example of using caregiving, and in this case motherhood, storylines strategically to encourage people to "feel" climate change on a personal level.

Other participants talked about ways their US-based environmental organizations think about connecting with mothers as a group. Sarah argued that they needed to bring in mothers who are active in their children's schools and to make them more aware that climate change is important for everyone. Janice explained that mothers are among the "unconventional spokespeople" about environmental issues because they are believable. She noted that people in meetings she attends will argue that "We need to get mothers involved. They care, you know, about the future of their children. Which is not to say that fathers don't, but women will talk more perhaps in that way." Rachel, likewise, referred to harnessing "the power of mothers":

> You can't really say bad things about mothers, right? A political bloc, or . . . I don't mean voting-wise, but just a group of people who could stand up and be respected. That comes to mind because I know about that . . . I've been in a couple other meetings with . . . women activists, women business leaders, . . . who are just asking the question "Can women as a group play a larger role in, or a different role, in climate solutions?"

Here, the belief that mothers are irreproachable means that they can "stand up and be respected." In societies across the world such beliefs might explain why motherhood discourses are prevalent in various kinds of activism, such as peace and anti-militarism (Carreon and Moghadam 2015; El-Bushra 2007; Gentry 2009). Movements that challenge a fundamental element of society (e.g., capitalism or militarism) can be "softened" through presenting it as something nonthreatening (Logsdon-Conradsen 2011). The same holds for appealing to femininity and motherhood discourses in particular in order to "sell" a wide group of people on climate change action. For instance,

Haley mentioned a female civil engineer she knows who wound up in the solar industry:

> She has kind of embraced this role as a "solar mom." I know, it's funny . . . because the first time I heard that I thought that was kind of offensive cause it's a pigeonhole type term. Like, "Why would you want to be a solar mom?" It is representative of this more welcoming attitude that the solar advocacy, or advocacy in general, does have toward women and mothers.

Although the solar industry's desire to seem welcoming can be viewed as a positive strategic move, the strategy might reinforce gender norms about who is most likely to perform care roles, a topic taken up in the next section.[8]

Interviewees from the nonprofit sector outlined other strategies they employ to soften their message about climate change, including refraining from using the term "climate change." Some told me they frame their message as being about children's health rather than climate change because this is seen as less threatening. Sharon is an environmental nonprofit worker in the United States and described the issue this way:

> So [my organization's] whole thing is really climate change, I mean we are trying to fight that, but because we work in the Southeast [United States] we don't call it "climate change" all the time because that turns a lot of people's ears off real fast, especially in the utilities sector. There are a lot of people holding on to the false notion that climate change is fake. . . . We focus more on the health benefits of decreasing carbon-intense resources. So, for example, not only are coal plants emitting carbon, but there's a bunch of other bad things both in the air and water. And so, a lot of time we will partner with health-based advocacy groups to talk about, you know, closing this coal plant will clean up the air for asthma kids.

Her remarks illustrate the strategic necessity of using frames that will both resonate with her organization's intended audience and avoid turning them off from the topic of clean energy. One tactic has been to partner with some of the parent-themed organizations discussed above to frame climate change mitigation as a children's public health issue.

In sum, the women-as-caregivers discourse contains storylines that focus on care in multiple contexts. Participants used the discourse to highlight women's roles of providing care for communities, as well as care within their own families. They discussed ways that women are taught to provide care, as well as whether women are instinctively nurturing. They reflected on their own identity as caregivers, or potential caregivers, and how climate change has shaped their identities. At the same time, they pondered how parenting influences climate change action. The discourse contains storylines that depict caregiving and motherhood as part of what influences how we understand climate change, how we act to address climate change, and even how we frame climate change to others in both personal and professional capacities.

IMPLICATIONS OF THE WOMEN-AS-CAREGIVERS DISCOURSE

What is climate change in the women-as-caregivers discourse? Climate change is a phenomenon that influences daily lives and experiences—including parenting and other care roles. It is a force that makes care work more difficult. It makes people sick. It makes resources more costly or otherwise difficult to acquire. Climate change is also a force that influences perceptions and mental health. It is noticeable when we provide care. It expands time horizons by highlighting connections to future generations. It results in an emotional toll as humans grieve for what will be lost to future generations and worry about what harm will befall this one. But climate change is also a space for advocacy and action to ensure that children and communities are as healthy and safe as they can be. Across the discourse, climate change is a phenomenon that shapes labor as well as perspective. It is not only an external, empirical fact but also a deeply personal force.

The international community has been grappling with how to think about women and/in environmentalism at least since the early 1990s. The 1992 United Nations Conference on Environment and Development (UNCED), or Rio Earth Summit, resulted in the international community adopting the concept of sustainable development to address environmental change while simultaneously recognizing a goal of economic development, particularly for the global South. The widely read accompanying document, *Agenda 21*, set out a plan of action for how states and other actors could (and should) make sustainable development a reality. The text specifically included a call for greater inclusion of marginalized groups, including women, youth, and indigenous communities. These groups were portrayed as living or being closer to the natural environment and therefore instrumental to pursuing environmental sustainability. Women in particular were depicted as reproductive mothers with a high degree of vulnerability and a strong capacity for environmental care (Bretherton 2003). Emma Foster (2017, 222) explains that in documents such as *Agenda 21*, "Women are constructed as closer to nature and the body—and separate from men, culture, and the mind—and therefore it follows that the environment is an obvious policy domain for women's decision-making 'abilities.'" In these treatments, solutions "to environmental problems were seen as requiring a more 'naturally caring' individual who had a greater investment in saving the planet for future generations due to her investment in child-bearing and child-rearing."

Where are the women in this discourse? They are in the global North and global South—in homes preparing food and caring for sick children. They are in communities working on climate change initiatives or engaging in gendered labor. Who are the women? They are cisgendered women whose biological relationship to care is not often challenged by their societies. They are caregivers and those who are concerned about the fate of their family, community, and planet. Their care roles motivate them to be active in climate change or sustainability initiatives. While some depictions of women in the discourse cast them as instinctively "nurturing," there was some evidence of self-reflection on the part of some participants about the socially constructed aspect of women's care roles. Nicola, an environmental nonprofit worker in the United States, stopped herself in the middle of the interview and said that she did not want to "put women into traditional gender roles" by talking about motherhood and care. She went on to explain that she understands the complexities surrounding this role or identity because it is a mix of social construction (we are socialized to think of women as the primary caregivers) and lived experience for many women. Sharon, also a US-based nonprofit worker, had a similar approach: I always come back to the woman as caregiver, even though that's such a stereotype and I don't want to at all say that men aren't caregivers, but I think about historically who has been caregivers. . . . I think about the women that will still be in charge of the kids and still be trying to figure out what to do with their family. Sharon recognized that she is using a stereotype and acknowledged that men also play caregiving roles. Yet she argued that past and current gendered divisions of labor often result in women having a lived experience of providing care.

While there is often pushback against homogenized ideas of women as nurturing or caregivers, it is also crucial to focus more on why men tend *not* to be expected to be nurturing, at least not in the same ways as women. Some interviewees raised the question of whether having more people "care" about the environment and other humans could potentially be a path forward to addressing climate change. Darcy, an academic in the United States, connected this to gender roles specifically, saying "men are socialized in society surely to have less sensitivity or care for an exploitable other. So, I think that that surely plays a huge explanatory role in why we're in the crisis that we're in right now." In this view, a lack of care for others shapes behavior in ways that are detrimental to the environment. This sentiment is also expressed by Shea, an environmental attorney in the United States, who observed:

> I hate to make nurturing into the feminine thing because it's so traditional and so stereotypical and is not true for a lot of women, but there's this nurturing aspect of, I think this nurturing piece is missing. I think one reason folks are detached from the issue of climate change is because they actually can't relate to it. Folks want to care about the air they breathe, this or that plant species, but they can't really. The language we use does not encourage this kind of development of a relationship with the environment. It encourages mainly "here's a problem, now we have to fix it" kind of attitude.

She highlighted that nurturing and care are two concepts that necessarily connect with people's attitudes on climate change and relationship with the environment. For both Shea and Darcy, societal attitudes with regard to care shape whether and how individuals are likely to act to address climate change. While Shea did not specifically talk about men or masculinity, Darcy indicated that the question of who cares and who does not is socially conditioned. It is not about who is most capable of providing care, but rather who is expected to do so. It is likely that the more people are encouraged to think outside of themselves—of those who are vulnerable to climate change, of species that are harmed by human activities—the more they would act to avoid the worst of climate change from occurring. In this view, caregiving is not some feminized task that women must think past, but rather it is a powerful reframing of environmental concerns that might result in sustainability and justice. In this way, a critical climate change and caregiving discourse can allow for the language and space to think about responsibilities and sustainability in meaningful and transformational ways (Whyte and Cuomo 2016). There are potential opportunities for utilizing expansive care discourses beyond the existing strategic use of maternalist storylines to reflect on climate change, a theme that will be expanded in the concluding chapter.

While the women-as-vulnerable discourse described in chapter 3 was almost exclusively used to think about women "in the developing world," participants used caregiver/motherhood storylines to reflect on their own experiences, those of women around them, and global patterns of gendered caregiving. Providing care or feeling a responsibility to care were depicted as common to women common across societies. As mentioned previously, participants provided different explanations for why women tend to provide care, but they conceived this as a phenomenon that transcends North/South boundaries. For these reasons, it is important to recognize the fact that motherhood/care discourse continues to be used to describe women's connection to environmental issues. The reasons for this are likely numerous and complex. The women whom I interviewed are products of societies that still tend to associate women with care. They are living and working in societies where women still have an unequal care burden relative to men (Bittman et al. 2003; Coffey et al. 2020; Schaeffer 2019) and where that burden is expected to grow heavier because of climate change impacts on disease and resource availability.

What are some of the obstacles as well as opportunities associated with framing women's involvement in climate change and other environmental issues through a specific focus on motherhood or caregiving? In terms of obstacles, feminist critiques point to the ways that environmental discourses that build on motherhood images are rife with homogenization and are quite likely to narrow understandings of environmental issues in ways that are detrimental to goals of both sustainability and justice (Bretherton 2003;

Butler 2017; MacGregor 2006; Perkins 2012 Sandilands 1999; Seymour 2013). Lois Gibbs, for instance, is one of the most recognizable names in the history of the US environmental justice movement. In 1978, she discovered that her son's school and the surrounding housing development was built on top of a toxic waste dump. She became a key leader in efforts to relocate the people of Love Canal who were impacted by toxic waste. Her efforts, along with the mostly female other members of the Love Canal Homeowners Association, helped to both push for stronger commitments to sustainability and environmental justice at the US Environmental Protection Agency, as well as raise awareness and action to address toxic exposure in homes. Despite her groundbreaking activism in this area, the media and public discourse centered on her role as a mother. "The perception was that her environmental activism was motivated primarily by her concerns about the contamination of the house and related threats to her children's health" (Sze 2017, 162). Media reports deemed her a "hysterical housewife" (Blum 2008; Seager 1996). These depictions ignored her agency as a person and focused solely on her position as a mother, while simultaneously repeating a "women are overly emotional" stereotype to belittle her actions (Perkins 2012; Sze 2017). Similarly, one participant I spoke to, Linda, explained that some of the politicians she interacts with treat her dismissively during the course of her work on child's health advocacy in the United States. She feels the need to convince them that she "isn't just a bored housewife who is doing this," but that she understands the science behind it and is making sound arguments about policy. The potential for the women-as-caregivers discourse to feed into tropes about women as overly emotional mothers necessitates thinking critically about care and environmental change frames.

At the same time, queer ecology argues that environmentalism's reprocentricity, cissexism, and heteronormativity result in a limited and potentially damaging conceptualization of care—both for people and the planet (Butler 2017; Seymour 2013). Scholars highlight the overwhelming concern for "future generations" in mainstream environmentalism as evidence of this reprocentricity wherein sustainability becomes linked to (re)productive potential (Seymour 2013). The public health crises that are often used to raise awareness of unsustainability are those that threaten the lives of mothers and children (Sturgeon 2008). Health concerns for communities regarded as less sympathetic (groups you can say bad things about) are typically missing in attempts to get people and policymakers to care about environmental issues.

An additional problem is that focusing on women as child bearers can result in them being seen as, at least partially, responsible for environmental change. The IPCC (2014b, 4), for instance, has specifically mentioned population growth as contributing to climate change.[9] There is a great deal of scholarship outlining the various instances of states enacting population control measures on women's bodies, particularly marginalized women (Cripps 2015; Hartmann 2010; Nelson 2003; Sasser 2018). For example, it is estimated that in the United States in the late 1960s and early 1970s thousands of women underwent forced sterilization. A large number of these women were Black, Native American, and Latina (Nelson 2003). While the idea of states addressing climate change by curtailing women's reproductive freedoms might seem a bit extreme to some of us, this has been the historical reality of marginalized communities when their society becomes convinced that population reduction is necessary. In a less extreme example, maternalist discourses can also exert additional pressure on women's reproductive choices at a personal level (Detraz 2021). Linking women with motherhood may result in pushing women to alter an important life choice in the name of sustainability. Rarely is the decision to become a father cast in the same terms.

One major critique of motherhood or caregiver frames is that they essentialize the category of "women" into a simplistic, homogenous entity. While participants mostly avoided conflating the positions of women and of children, there were instances of essentialization across the interviews. Some of this language was akin to early academic and activist versions of ecofeminism. The term "ecofeminism" traces back to 1974, when French feminist Françoise d'Eaubonne used the word *ecoféminisme* to highlight women's action as necessary to save the planet. In later years, many scholars and activists use the term "ecofeminist" to refer to their efforts to link feminism and ecology. While ecofeminism covers many approaches to connecting feminist concerns and environmental concerns (Mies and Shiva 1993; Warren 1997), initial versions of the concept have been criticized for putting forth an essentialized depiction of women as caregivers of the planet (Sandilands 1999).

On the other hand, it would be a mistake to dismiss caregiver storylines as always detrimental. I was fairly surprised by the frequency of the women-as-caregivers discourse in the interviews. I went into this project with a strong understanding of the feminist critiques of essentialist versions of ecofeminism, motherhood environmentalism, and eco-maternalism. However, some version of a caregiver/motherhood storyline kept appearing as I interviewed more and more women. Women were described as "nurturing," "the carers," and "the first point of contact in the home." While I agree with those critical voices who reject homogenized depictions of women and care (Detraz 2017b), I also find it interesting and important that a women-as-caregivers discourse was used by women working on climate change issues to understand their own position as well as general concerns about climate change. Many interviewees shared very personal reflections about what being a mother or caregiver meant for their feelings about and work on climate change. The women-as-caregivers discourse gave them the storylines to share this part of their identity and experience. What is noteworthy is that many women see their role as a mother, or the role of caregiver in general, as a vital component of their existence. And what becomes significant is how this discourse might be influencing current conceptualizations of climate change. Scholars and practitioners must grapple with how to avoid essentialization and representing women in ways that are likely to reduce them to "hysterical housewives" with little to say about public concerns, while also recognizing that caregiving and motherhood form an intimate part of many people's identities and relationships to environmental concerns. It is undoubtedly true that there are considerable problems with dichotomous understandings of women as emotional and caring on one hand and men as rational and stoic on the other, but it is also true that all of the interviewees who used a motherhood/care discourse also used other discourses that signal women's expertise, ingenuity, resilience, and agency. We turn to these storylines next.

5 WOMEN AS KNOWLEDGEABLE: EXPERTISE AND COMPETENCE IN CLIMATE CHANGE

Glenda met me for our interview at a busy coffee shop in London. She bustled in with her bike and helmet, ready to talk about any topic I threw at her. Her legal training shone through as she rattled off names, dates, and specific environmental cases effortlessly. She traced the start of her environmental work to 1992, when she helped start a bicycle group in her community. She recalled the enthusiasm about environmental issues in the UK and elsewhere that year as it coincided with Rio Earth Summit. Since that time, she has continued community-level action on environmental issues as well as working as an environmental lawyer. She has done a great deal of legal work on fracking, among other environmental issues, and has insisted on using a climate change lens to frame arguments against fracking. In the course of our conversation, she mentioned several prominent female figures who she thinks play an essential role in climate change. Focusing mostly on political and diplomatic expertise, she mentioned female politicians, women who lead environmental organizations, and women playing a role in global negotiations. In her view, "Female leadership is very important. I mean if you look at Christiana Figueres and what she's doing. She's in the, she's almost like the beacon in it." For Glenda, women are active and knowledgeable participants in climate change work.

In a book on women and climate change, it is perhaps unsurprising to find discussions of the discourses examined in the two previous chapters, women-as-vulnerable and women-as-caregivers. Women's vulnerability, particularly associated with care roles, is firmly established in the popular imagination as well as global policy debates. We now turn to quite a different perspective on the position of women. As illustrated by Glenda's story, participants regularly considered themselves and other women to possess valuable knowledge that is essential to addressing or understanding this global problem. This chapter explores the *women-as-knowledgeable* discourse.[1] It first describes perceptions of women as well-informed about climate change. This knowledge is seen to manifest in women being less likely to engage in climate skepticism, and to educate others about climate change. The chapter then examines some of the roles that interviewees identified for knowledgeable women. These roles specifically include climate change leader/diplomat, climate change educators, and climate change researchers. I also analyze some instances in which participants had to push back when their expertise was called into question. I close with some implications of the women-as-knowledgeable discourse.

KNOWLEDGE, EXPERTISE, AND CLIMATE CHANGE

Many scholars have studied the links between knowledge, expertise, and climate change (Pettenger 2007). Research covers such topics as the role of local knowledge (Naess 2013; Reyes-García et al. 2016) or indigenous knowledge (Alexander et al. 2011; Green and Raygorodetsky 2010) in climate change mitigation and adaptation. This work is often motivated by the argument that climate change debates have been dominated by elites and that effectiveness or equity goals or both require this to change. Additionally, research examines linkages between climate change knowledge and belief in climate change occurring (Guy et al. 2014),[2] connections between climate science and policy (Meyer 2011), and the dominance of the global North in perceptions of expertise and publications about climate change (Pasgaard and Strange 2013; Pasgaard et al. 2015).

There is also literature that examines the specific place of gender in climate change knowledge or expertise, including feminist work exploring how power relations shape our knowledge and assumptions in fields such as climate science (Israel and Sachs 2012; Moosa and Tuana 2014; Tuana 2013) and climate economics (Nelson 2007). This work examines how climate science and climate economics center concepts such as "value-neutrality" and "objectivity" in ways that reinforce existing gendered power dynamics as well as limit our approaches to understanding and addressing climate change (Moosa and Tuana 2014). Other work examines how gendered divisions of labor produce specific forms of climate change knowledge (Agarwal 2009; Merchant 1995; Warren 1997). These studies often include an argument that this gendered climate change knowledge necessitates women's inclusion in climate change decision-making (Terry 2009). For instance, Trish Glazebrook (2011) uses the case of women subsistence farmers in northeast Ghana to argue that women tend to play specific social roles that render them uniquely vulnerable to climate change while also affording them specific information relevant to climate change policy.[3] She claims that "attention to [women's'] situations in responding to climate change must focus not only on their experiences as victims, but also on the contributions they can make to the climate struggle as resilient and expert actors" (Glazerbrook 2011, 769). There is thus both a justice and a sustainability component to the argument: it is just to include women because they are particularly vulnerable, and it is practical and helpful to include women because it allows for fuller climate change knowledge that is important to mitigation and/or adaptation efforts. At the same time, some scholarship warns of the dangers of coopting women's environmental or climate change knowledge (Dey, Singh, and Gupta 2018; Dove 2006; Federici 2009; Sapra 2009). These studies call attention to women's environmental knowledge being treated as merely an input in climate change policymaking without women also being recognized as stakeholders and experts.

Anna Kaijser and Annica Kronsell (2014, 419) expand on this idea by offering an intersectional analysis of climate change:

> Intersectionality can generate alternative knowledge crucial in the formulation of more effective and legitimate climate strategies. Intersectional analysis has a normative agenda, as feminist and critical theories generally do. It is related to the feminist epistemological position that regards knowledge as derived from social practice. This way, intersectionality also highlights new linkages and positions that can facilitate alliances between voices that are usually marginalised in the dominant climate agenda.

This perspective reflects on ways that multiple positions intersect to influence climate change knowledge. Within views like these there is no such thing as a generic "women's climate change knowledge."

Another strand of the literature looks at connections among gender, climate change knowledge, and climate denial (Anshelm and Hultman 2014; McCright and Dunlap 2011; Nagel 2016; Smith and Leiserowitz 2013). Existing empirical work lends strong support to the idea of gendered patterns of climate skepticism or denial. For instance, Aaron McCright and Riley Dunlap (2011) find that in the United States white, conservative males are more likely than other adults to express climate change denial. In the same study, they also report that those white conservative males who positively assess their own understanding of climate change express an even greater degree of climate skepticism than others. This denotes that while these men are highly confident about their own levels of knowledge, this confidence is misplaced. These conservative men underestimate the likelihood of climate change occurring, the causes behind it, and the level of consensus within the scientific community about these points. Another study conducted with a sample from the United States finds that whereas evangelical Christians in general tend to perceive climate change as less of a risk than non-Evangelicals, Evangelical women tend to assess climate change as a risk more frequently than men within this group (Smith and Leiserowitz 2013).[4] The finding that men, and particularly conservative men,[5] are overrepresented among those who express climate denial positions holds for other research using data from Brazil, New Zealand, Norway, Sweden, and the United Kingdom (Anshelm and Hultman 2014; Jylhä et al. 2016; Krange, Kaltenborn, and Hultman 2019). These are all countries that have seen significant shifts in economic sectors over the past few decades, including the decline of manufacturing and shifts away from fossil fuel. Paul Pulé and Martin Hultman (2019) argue that the "industrial breadwinner masculinity" present in these states responds strongly against forces that appear to threaten its privileged position.[6] This includes gendered patterns of climate denial.

While there is a committed segment of the global population that persists in climate denial, trends indicate that this may be a shrinking group (Poushter and Huang 2019).[7] Climate skepticism has declined in the United

States and Europe, regions where most interviewees live. A 2018 study on climate change attitudes in the United States found that 73 percent of those surveyed think climate change is occurring (Leiserowitz et al. 2018).[8] This represents an increase of ten percentage points since a March 2015 study. Only 14 percent of the respondents think global warming is not occurring. Likewise, a 2017 Eurobarometer report found that 92 percent of the respondents believe that climate change is a "serious problem," and 74 percent of them consider that climate change is a "very serious problem" (European Commission 2017). The latter number is up from 69 percent in the previous 2015 Eurobarometer poll. While responses vary for citizens in different parts of the European Union, some of these regional differences have narrowed (European Commission 2017).[9] There are also gendered patterns in the responses. Women are slightly more likely to say that climate change is a very serious problem than men (76 percent to 72 percent) and are slightly more likely to indicate that they have taken personal action to fight climate change (51 percent to 48 percent) (European Commission 2017). Several of the themes and findings from the academic literature on knowledge and climate change also appear in the women-as-knowledgeable discourse. The following sections will specifically explain the study participants' views about how gender plays a role in acquiring and assessing climate change knowledge.

Women's Knowledge and Climate Skepticism

Two key storylines that fit together in the women-as-knowledgeable discourse are the idea of women-as-knowledgeable-about-climate-change and women-as-concerned-about-climate-change. Interviewees described women knowing a great deal about climate change in a general sense as well as having unique comprehension of climate change processes and impacts. Janice, who works in the nonprofit sector in the United States, reflected on women's attitudes about climate change and how women and men might diverge in their approach to the issue. She noted that "especially when you talk about denial, the manufactured denial which has largely been, you know at behest of the fossil fuel companies, what we see in the polling that we look at is that the women are much less susceptible to those messages than the men are." This perspective was shared by Kristy, who works for a climate research group at a university in the United States. When I asked for her initial impression of the phrase "gender and climate change," she responded:

> Retired white males . . . What do I mean by that? Any time I've given public presentations it is usually the retired white males that are the grumpy ones who are going to make some claim about sunspots, or yeah, climate change being "natural." That sounds so calloused, but I mean it's true. But to their credit I will give them credit for showing up. I mean there a lot who wouldn't . . . but it is kind of funny because you can usually pick them out.

In her experience, men—and particularly older white men—were more likely than women to express climate skepticism at her lectures.[10] Their references to "sunspots" or naturally occurring patterns of climate warming and cooling might indicate that these people tend to feel more confident in their own expertise on matters of climate change, although in fact, these common climate skeptic talking points have been largely debunked in scientific scholarship (IPCC 2014b). This view that women are less represented among the population of climate skeptics was repeated by participants in the academic and nonprofit spheres. They pointed to evidence from polling data, experience in their classrooms, interactions with members of the public at speaking engagements or public outreach events, or conversations with people in their lives of women being less likely to publicly express climate skepticism.

As discussed above, existing empirical work lends strong support to the idea of gendered patterns of climate skepticism or denial (McCright and Dunlap 2011; Smith and Leiserowitz 2013). Many participants mentioned that, in their experience, women tend to voice greater concern about climate change. In the words of Sharon, a clean energy specialist in the United States,

> I see in my reality that there are more women usually on one side of the debate and men on the other. Women on the "Yes climate change is real, and we need to act now." . . . The first thing I think of is women are better suited . . . traditionally . . . to jump into this belief, to join the fight, so to speak.

For Sharon, you cannot have climate change action without first having some awareness of the basics, as well as the scale of the problem. This is one way in which interviewees established a connection between women's knowledge and concern about climate change. Another link between women's knowledge and environmental concern is that of mothers passing on climate change knowledge to their families. This mothers-as-educators storyline emerged in multiple interviews and serves as a bridge between the women-as-knowledgeable discourse and women-as-caregivers discourse explored in the previous chapter. German nonprofit worker Lyra, for example, posited that organizations need to think long term about climate mitigation and adaptation projects to better plan for the future. She said that women need to be active participants of community-based projects because they will be the ones who can pass the knowledge on to their kids. Swati, an environmental nonprofit worker in the United States, highlighted these links, noting "the role of women as, like, storytellers in their families, and passing along those stories to the next generation of how things used to be, and kind of . . . seeing firsthand." She likened this process to indigenous communities teaching their members to think several generations into the future. She wondered, "If we adopted that, how different would our policies be, and our choices be?" Swati reflected on how her own views were shaped by her upbringing:

> I'm an immigrant to this country, my parents and I moved here from India when I was really young, like three or four, so I was always raised with that environmental ethic of conservation and our own conservation of our consumption. . . . So, we still always treated it as, "You turn off the lights!" . . . I had some colleagues stay over last week, and they just keep the lights on. They leave the room, and they keep the lights on. I'm like, "What are you doing?" Yeah, and it's just, like wow, when that gets instilled in you as a little kid like that's a habit . . . that you can't break.

Swati recognized her parents' lasting influence on her attitudes and practices related to the environment. To this point, Sina claimed that she specifically likes to do outreach to mothers with her US-based nonprofit since they bring information about climate change home to their families. Additionally, US-based Elane mentioned that women have the capacity to teach their children about climate change and therefore help usher in necessary change. Sina and Elane both argued that this educating role is something very "powerful." Since knowledge production and understanding of climate change are socially conditioned, early exposure to pro-environmental information can potentially shape behavior for future generations, an argument that has some empirical support (Matthies, Selge, and Klöckner 2012; Matthies and Wallis 2015).

While a few interviewees suggested that they see women as more likely to care about environmental change and climate change in particular, others claimed that this does not necessarily translate into changes in policy or other types of action. Vicky works for an environmental organization in the UK, and she recounted seeing a study that finds that "young women in the UK are the demographic who are the most likely to say they care about climate change, but also the least likely to ever talk about it. And they think that's because women are more aware of what's socially acceptable around them and they pick up on the fact that it's not socially acceptable to talk about climate change." This unwillingness to talk about climate change publicly for fear of backlash speaks to the social and politicized nature of climate change assessment and knowledge. While exposure to pro-environmental messages might influence children's behavior, those same people are subjected to additional socialization that gives them clues about what behavior will be rewarded or punished.[11]

Where Does Environmental Knowledge Come From?

One storyline in the women-as-knowledgeable discourse is the idea that women have specific forms of knowledge. Participants often linked this to gendered divisions of labor in households or communities. Vanessa reflected on this in terms of who takes part in different types of activities hosted by her Scottish environmental nonprofit. She mentioned that men tend to get involved in energy or transportation projects, while women are more likely to participate in recycling or food waste projects. When I asked her why she thinks that might be the case, she said:

> I mean, I hope it's not because people are conforming to traditional gender stereotypes, but I mean, it might be that women are more involved in the cooking and therefore more, have more knowledge of the waste that goes along . . . things like that. And I really don't know . . . I guess there is still a bit of that—a predominance of men in . . . the energy-type sector, the engineering . . . that side of things. Maybe there are more men with more knowledge of that area who therefore transfer that over to their . . . that kind of work.

In her view, gendered divisions of household labor might influence the kinds of issues that one is most aware of. With women perhaps playing an oversized role in food shopping and preparation in many families, the issue of food waste might be more readily on their radar.

As discussed in chapter 3, women are frequently cast in the role of "worker" within the interviews as well as general discussions of climate change in the media, policy community, and academics (Arora-Jonsson 2011). Multiple interviewees specifically identified women's work in agriculture as well as food and water collection as contributing to their climate change knowledge and their ability to understand when climate change impacts are occurring. They commented on gendered patterns of men engaging in growing crops for profit, for example, and women engaging in subsistence farming. Kit, for one, brought up her experience studying resource management in Nepal to point out these gendered differences:

> Women harvest leaf litter and firewood. Men cut timber that takes them to different areas of the forest. So, they have a different understanding of different areas of the forest. . . . The women's is kind of more daily and local. So, the way that they would experience any impacts from climate change on the forest would be different to the way men do.

Eva, a nonprofit worker in Germany, made a similar comment about women's environmental knowledge in places where they are responsible for finding water or biomass for fuel:

> It's the women who know best. They are very familiar with the changing environment and the dangers that it might entail. So, any effective solution should be developed in close coordination with local women, I would say.

Whereas Kit drew specifically from her fieldwork, and Eva referred to a general "local women's" perspective, both highlighted gendered divisions of labor and women's knowledge in rural areas of the global South[12] in ways that illustrate connections between the women-as-knowledgeable and women-as-vulnerable discourses. Interviewees argued that gendered divisions of labor for rural women in the global South afford them specific forms of knowledge about environmental processes, but this position also makes them uniquely vulnerable to the impacts of environmental change.

However, a few participants worried that a focus on women's resource knowledge can be incorporated into climate change debates in ways that are not necessarily helpful for women's full participation in the conversation. Again, when interviewees mentioned women as resource users, they were typically referring to rural women in the global South, who typically lack a strong voice in political decision-making and are rarely regarded in climate change policy spheres as "experts." Rather, these women tend to be viewed as having specific pieces of environmental knowledge, which does not qualify them for the same position at the decision-making table as the experts (MacGregor 2006). Darcy, a US-based scholar, expressed frustration with the trend of strategically "using" women's knowledge without necessarily giving them the same status as authorities on environmental change or viewing them as equal participants. She brought up women "specifically being called upon to combat climate change as mothers of the earth and certain gendered, patriarchal norms that they can play into how we turn to women to combat climate change." Marie, a German nonprofit worker, also pushed back against this instrumental approach to women's knowledge by asserting that they should have a role in finding solutions. "Well, I think that if you try to solve a problem, you should definitely get those who are, who are affected by the problem have a seat at the table. So, in this regard, women probably have knowledge of solutions or should be involved at least [to] an equal extent as men into solutions to climate change."[13] In this view, women should be included in decision-making since they are both well-informed and disproportionately impacted by the problem. This concern about the global community using rural women in the global South, or other marginalized groups, as sources of information rather than as stakeholders also appears in academic work (Dey, Singh, and Gupta 2018; Dove 2006; Federici 2009; Sapra 2009). Hence, women's participation is regarded as necessary to achieve sustainability goals as well as justice goals.

ROLES ASSOCIATED WITH KNOWLEDGEABLE WOMEN

While the preceding section discussed women's climate change knowledge in general terms, the interviews yielded a picture of specific jobs or roles related to expertise that women assume in climate change spaces. The image of knowledgeable women took several forms, including climate change leaders or diplomats, researchers creating data and information, and experts providing knowledge to various actors. It is important to note that these roles are not mutually exclusive. In fact, interviewees identified multiple, simultaneous roles that relate to climate change expertise. Some took on a defiant tone when identifying specific tasks that women play because of their position as experts on climate change—usually after an observation that many climate change spaces continue to be dominated by men. For them, the fact that women play these expert roles and possess knowledge and valuable skills means that women are an essential part of the climate change realm and should be respected accordingly.

Climate Change Leaders

Participants referred to women as "climate change leaders" or "effective negotiators" multiple times.[14] "Across the globe, when women are . . . empowered politically, we see big changes" was US-based nonprofit worker Gwen's comment on women's position in climate change. Additionally, German nonprofit worker Marie's initial response to the phrase "gender and climate change" was that

> As a field, climate change policy has shown more women leaders than in other, in other fields of policy, let's say, in general. Especially within the climate change negotiations. Christiana Figueres is of course the big leader of all, but I think [women's participation] was quite visible at the negotiations in Paris. . . . There are quite a lot of women that are involved in solving the issue and [they] tend to play more of a leading role than in other sectors.

Like Marie, other interviewees specifically identified Christiana Figueres as a woman who is widely recognized for her prominent role in global efforts to combat climate change.[15] She was the Executive Secretary of the United Nations Framework Convention on Climate Change (UNFCCC) from 2010 to 2016. She is often credited with playing an essential role in getting the international community to finally sign on to the Paris Agreement.[16] Numerous outlets praise her proficiency in climate change diplomacy. She is called a "world authority" on global climate change (World Resources Institute 2018), as well as "a widely published author on the design of climate solutions" and "a frequent adviser to the private sector" (United Nations Climate Change

2019). These accolades underscore a unique skill set and wealth of knowledge pertaining to climate change negotiations. Participants typically had a note of pride in their voice when they mentioned Christiana Figueres. They seemed to regard her accomplishments as evidence of women's potential and proof of their ability to play a key role in climate change.

Brooke, a scholar in the United States, is an example of someone who mentioned Christiana Figueres in her initial reaction to the phrase "gender and climate change." She went on to discuss "the role of women as negotiators and leaders in the climate arena" and specifically identified women as the ones "who basically gave us the Paris Agreement." Lyra also referenced women's participation in climate negotiations.[17] She works with delegates to the United Nations Framework Convention on Climate Change (UNFCCC) from small island states in her role at a German climate nonprofit. She reflected on the strong representation of women in the groups that she works with, specifically on the finance team. Lyra also mentioned their coordinator, who is an ambassador to the UN:

> [She is a] really experienced negotiator and she coordinates the whole group . . . and a lot of young women from ministries of other countries, or like me from research organizations in the team. And I think for all of us it's very cool and also important to see what she does and how she's doing it. And I can even see just from being in this team for three years, how a lot of the young women in the beginning who were very, very shy have now been empowered . . . kind of . . . through working with her in this team where she also gives them the opportunity to speak or to prepare a certain issue.

Lyra's comments speak to women's knowledge and position, as well as what this means for other women in a group. She argued that the presence of an experienced negotiator allows others to learn from her. This is also facilitated by the fact that she gives them opportunities to grow. Lyra referred to young women being "empowered" by this process of mentorship and participation. However, for "empowerment" to occur, women must be present in the first place.

Women's participation in global climate change negotiations has been increasing over the time, yet it is still relatively low compared to the stated UN goal of gender balance. The international community recognized the need to improve women's participation in negotiations in 2001 at the

UNFCCC Conference of the Parties (COP) 7 in Marrakech. There was a follow up in 2012 through an additional decision on promoting gender balance at COP 18 in Doha, Qatar. The figures for women's participation at COP 24 in 2018 and associated meetings lagged well beyond a goal of gender parity, with women making up just 38 percent of party delegations and 27 percent of heads of delegations (UNFCCC 2019).[18] While these figures represent a modest increase from the previous year, they reveal that women remain underrepresented in UNFCCC gatherings. However, women have played a large role in the Women and Gender Constituency (WGC) at the UNFCCC. This group provides avenues for civil society and nongovernmental organizations (NGOs) working on gender justice, environmental protection, or both, to shape the UNFCCC and annual meetings. It has been recognized as one of the nine official observer constituencies within the UNFCCC since 2011. Even though women's participation as delegates in climate change negotiations continues to lag behind men's, some interviewees identified this as an area in which women's climate expertise is on display and where their involvement is necessary. None remarked on the other forms of women's participation in climate change negotiations like the WGC. This means that while they recognized women's expertise and underrepresentation in "official" channels, they were not aware or else failed to mention these other forms of expertise and participation.

Commenting on women's knowledge and political adeptness, Gwen, who works for an environmental nonprofit in the United States noted that women in positions of political power are more likely to achieve legislative wins on climate change, often through compromise.[19] This women-as-leaders storyline manifested when participants suggested that women are particularly skilled at incorporating multiple voices into the policymaking process, as well as thinking broadly about complex issues such as climate change. Interviewees argued that women are more likely to listen to other perspectives, are more likely to be among those working to diversify the environmental movement, and are better at including marginalized peoples, including other women. These characteristics speak to their ability to be effective in achieving climate change action, particularly in ways that are also sensitive to social justice goals.

Some participants specifically mentioned women's ability to view climate change "broadly" or "holistically." An example comes from Brenda, who works for an environmental nonprofit in the United States:

> I think that the solutions that we need are going to come from a different place and a different type of thinking than that type of thinking that we have used for the last, you know, hundreds of years—since the Industrial Revolution. And that is a more traditionally feminine type of thinking. It is a slower, more holistic, more community focus, less profit driven, more labor, less technology . . . maybe. You know, using technology in a way that supports people as opposed to exploits people. And I think that's a traditionally gendered way of thinking.

Brenda went on to say that while holistic thinking is traditionally feminine, men are perfectly capable of viewing global issues this way. While there are men currently doing this work, we need large-scale changes in perception so that they feel comfortable and supported when they advocate for this kind of approach as well. Her comments indicate that women are accustomed to viewing the world in ways that are beneficial for climate change action. She regarded their perspective as valuable for driving the societal shifts necessary for achieving sustainability. However, existing scholarship indicates that the mere presence of women within institutions does not always correspond with progressive policies, a point elaborated on in the next chapter (Magnusdottir and Kronsell 2015).

Knowledge Creator

Participants also identified a strong role for women in the creation of knowledge about climate change, often by referencing women's work within the scientific community. One interviewee pointed out that "there are lots of women in science overall, including strong representation in fields like botany and marine biology where a lot of the work monitoring the effects of climate change on ecosystems is being done." Others mentioned that "the climate scientists that are out there now talking about the issues—Katherine Hayhoe, Heidi Cullen—these are women who are leaders in their fields." As in the case of Christiana Figueres discussed above, these mentions of high-profile women in climate science were accompanied by a tone of pride that women's position in their field is recognized.

Multiple participants identify as scientists and have training and/or are working in a natural science field. They frequently discussed women's capacity for research in their responses. For instance, Heather, a biologist/biogeographer who works at a university in the United States, thought that "women do fabulous science. We just haven't had the opportunity to do fabulous science for a very long time, but I don't think of [climate science] as something that is by nature masculine at all." For her, the ability for women to do great science is unquestioned. While she pointed out some of the challenges that women can face in natural sciences, she argued that these do not speak to women's capacity for scientific research.

Additionally, Jasmine, a US-based expert on plants and climate change, said that she decided to become a scientist in order to gain specific kinds of knowledge in her field, which facilitates her voice being valued. She argued that this knowledge makes her credible. Jasmine also indicated that her position in science allows her to recruit other women, particularly other women of color, into fields like hers. She explained that she thinks about this when she chooses students to work with, because while she has noticed more gender balance in her field throughout graduate school, she noted an underrepresentation of women of color. She said, "Any activity I do, I try to think 'how I can involve other underrepresented groups?'" Jasmine saw her membership in the scientific world as important not only for the research she does, but also for how it facilitates outreach to others.

Some storylines about women's expertise in science involved situating women's presence and experiences in their fields more broadly. Two commonly used storylines were women's-increased-participation-over-time and science-as-gender-neutral. Multiple interviewees observed that while scientific fields like theirs have long been male dominated, this is changing. Incoming cohorts have larger numbers of women than before. Constance noted:

> It seems that there are a lot more females than males typically working on climate change–related projects. . . . For my undergraduate institution that was the case. . . . A lot of the students that were in my atmospheric science department and were really interested and passionate about climate change and making a difference were mostly female. Which is funny when you look at the larger field of atmosphere science, there's such a small pool of females doing it.

Heather, another participant, mentioned, "We're at a funny time, I think, because it does seem like there's more women coming up now. And in terms of academia . . . assistant professor positions, things are changing." However, Heather went on to say, "And yet, most of the, sort of, loudest, most dominant, and most authoritative voices in the room are always still men—in many cases in atmospheric sciences. And the voices around climate change, and even the voices around dissent, and, like, doing something about [climate change]." Her reflections fit a common pattern among the interviewees, who noted that while women are increasingly being recognized for their proficiency in science, there is still quite a way to go before they attain the same level of prestige as their male peers.

Even as they recognized that several scientific fields remain male dominated, nearly all participants who work in these fields strongly believed that science is not a gendered process. This was largely in response to my asking them whether they view climate change as a masculine issue area. I chose to ask this question in order to introduce the concept of masculinity into our conversation, as so many interviewees focused almost exclusively on women in their responses. Scholars such as Sherilyn MacGregor claim that climate change in particular "has brought about a *masculinization* of environmentalism. Men dominate the issue at all levels, as scientific and economic experts, entrepreneurs, policy makers and spokespeople" (2009, 127–129). Additionally, she maintains that understandings of climate change and the accepted solutions to it are gendered in that "climate change has been presented not only as a largely scientific problem (one might say it has been *scientized*), but also as a threat to national and international security (i.e., it has been *securitized*)." She, along with many other feminist scholars, argues that science and security are "stereotypically masculinist discourses." This means that they are areas that have been dominated by particular ways of thinking that are associated with masculinity (Fox Keller 1995; Harding 1991, 1993). While they have also tended to be male dominated, it is more their general orientation and prevailing discourses that makes them masculinist.

While it is important to note that they were not reacting directly to MacGregor's full argument, some participants were quite unsympathetic to the idea that climate science could be conceptualized as masculine. Although multiple participants were receptive to the idea of reflecting on climate change as a masculine issue area, others thought it was "rubbish," "annoying," and "really rankles" them. April, a nonprofit worker, claimed that there are many women in Germany with the "same knowledge, expertise, experience, level of vocabulary and so on [as men]." For her, this meant that women are present and proficient in science. Likewise, Kate, a nonprofit worker in the United States, argued quite strongly that

> I've never felt science can be a masculine endeavor. And I think I can credit my father for that—because he is a scientist, and he always taught my sister and I both to handle things in a logical and analytical way. And I think it sounds like a gender cop-out to make it feel like it's a mutually exclusive issue. Like science versus ladies. Do you know what I mean? . . . Who knew I'd feel so strongly about that?

Kate's idea of science entailed the use of logic and analytical thinking in dealing with a given problem. She focused on a process rather than a field with messy boundaries and unequal distributions of power, including power among members but also power between humans and the natural world. This later picture is closer to what feminist scholars such as MacGregor describe. Kate's portrayal of science fits the science-as-gender-neutral storyline that understands the distribution of men and women in science as problematic, but not science itself. Feminist scholars have published a great deal on the links between gender, science, and technology. The contributions of feminist scholars to the philosophy of science have been immensely important for shedding light on how notions of masculinity and femininity infuse scientific study and practice (Fox Keller 1995; Harding 1991, 1993). This literature calls for recognizing that both gender and science are socially constructed categories. Yet the former is more often recognized than the latter as a constructed notion. According to Evelyn Fox Keller (1995, 4), "Science is the name we give to a set of practices and a body of knowledge delineated by a community, not simply defined by the exigencies of logical proof and experimental verification." She goes on to explain that "women, men, and science are created, together, out of a complex dynamic of interwoven cognitive, emotional, and social forces." Understandings of the boundaries of science, including what counts as science and who counts as a scientist, are not law. They emerge from a set of social negotiations that are often invisible and always in flux. The science-as-gender-neutral storyline is largely incompatible with this view of science. Those participants who used the storyline typically did not have a great deal of background in thinking about gender in their work. This is one of the reasons that understanding discourses is so important: it can potentially offer tools for how to reflect on gender in ways that some people working in climate change have rarely been expected to.

Knowledge Provider

While the scholars I spoke to identified themselves as having expertise in their particular fields, women's proficiency in academia is not always recognized at the same levels as men's. In my field of political science, for instance, research illustrates that women are consistently cited less than their male peers, even when accounting for other significant predictors of the difference (Maliniak, Powers, and Walter 2013; Mitchell, Lange, and Brus 2013).[20] There are organizations whose goal is to shift assumptions about where scholarly expertise comes from, such as the Women Also Know Stuff database in Political Science, which sets out "to promote and publicize the work and expertise of scholars in political science who identify as women." They explain that "implicit and explicit gender biases mean that women are often underrepresented as experts in the academy and in media" (Women Also Know Stuff 2019). The fact that people felt this database was necessary speaks to the enduring challenge that women face in the quest to be seen as knowledgeable. It would be unfathomable to have a "Men Know Stuff" database because that is already assumed.

Likewise, the group 500 Women Scientists was founded by a small group of women right after the 2016 US presidential election through an open letter stating a "commitment to speak up for science and for women, minorities, immigrants, people with disabilities, and LGBTQIA." They have since grown into a grassroots organization with over twenty thousand women from natural science and math fields and supporters from more than one hundred countries signing up in support of the organization. Their current mission is to "serve society by making science open, inclusive, and accessible" (500 Women Scientists 2019).

Both of these organizations serve to highlight women's expertise in the social sciences and sciences. In both cases, the motivation behind them is to show other communities what my interview participants already think—women know stuff. They particularly know stuff about climate change.

While interviewees largely referred to other women as "climate change leaders" in the realm of negotiations or diplomacy, they highlighted their own expertise in their role as researchers, consultants, or teachers. They work on many topics that relate to climate change within various disciplines. For these participants, climate change knowledge is acquired and passed on through multiple sources, especially through education, research, graduate mentoring, and teaching. At the beginning of each interview, I asked participants some basic information about how long they have worked on climate change and what kinds of tasks they perform in this area. This is often where interviewees indicated their own role as experts on some aspect of climate change, ranging from climate change ethics, to climate change governance, to the role of clouds in the climate system, to climate-related migration of trees, and beyond.

Some participants from the academic sector recounted ways that their expertise was recognized and sought out by different actors. For instance, Brooke, a US-based academic, was invited to contribute to a local-level climate change project because of her expertise in climate change governance. Additionally, US-based Candice frequently provides her expertise for a range of actors, including a children's environmental health organization and the US Navy. She has also testified before the US Congress. These examples illustrate that many interviewees not only acknowledge their own expertise but are also regarded as authorities in their field by other actors. In Candice's case, the US Navy, which many might not immediately associate with the climate change realm, asked her to consult with them multiple times based on her research on ice in the climate system, a topic of interest for those concerned about sea level rise and other impacts that have been linked to issues of national security (Center for Naval Analyses 2007).

More than one participant has done work as a "knowledge broker," or someone who works as an intermediary to translate information from one source into something usable for a different network. In these cases, the women translated climate science for policymakers or members of the general public. They used a particular skill set of distilling and reframing climate data, a task that requires expertise in both climate change and policymaking or public engagement. Candice argued that the public needs to understand climate change impacts beyond numbers. At the same time, this knowledge broker role might not be validated in academic spaces:

> What I do is societally relevant except it has to be translated into something that society needs. They don't particularly need any climate model unless somebody helps them . . . interpret it for what it means to society. And that's where, even though it probably won't win the respect of my peers, as I've always attempted to do, the route that I've been affiliated with anyway. . . . And so, I feel like I can contribute something unique and probably critical to success. I'd like to do that.

Candice saw her contribution as unique and essential for successfully engaging the public about climate change. She viewed herself as capable of providing this function, even though she felt she is not rewarded for this kind of activity in her career.

This knowledge broker role was also played in a less formal way. Constance, for example, started a "climate change communication" group with another student at her undergraduate institution in the United States to present climate science to middle and high school students. She wanted to present the science to them in an "unbiased way," giving them "just the facts." Other interviewees in the nonprofit sector told me that their goal is to arm people with knowledge about climate change so that they can make informed decisions as consumers or citizens. Others interact with policymakers through lobbying efforts. What is important to note here is that these women viewed themselves and their organization as possessing the climate change knowledge that they transferred to other actors.

In teaching and mentoring students, scholars convey climate change information as part of their routine duties. Sarah is a physical oceanographer who heads an interdisciplinary program on climate change at her university. As part of the program, she helps coordinate a yearly Graduate Climate Conference, which draws eighty students from across the United States. She referred to this as "the thing she is most proud of" in her work with the program and noted that this event "shows that we're creating broader, more intellectually open scientists." This knowledge dissemination activity was something that she was particularly excited to share with me. Additionally, Kit, an academic based in the UK, mentioned that she touches on gender and climate change connections in one of her courses that tends to contain large numbers of biology majors. She commented on the fact that students typically come into the course with an awareness of climate change, yet many are surprised by learning how it intersects with gender. The fact that she introduces new students to these issues each time she teaches the course means her expertise is being repeatedly distributed to a new population. In the course of her position as a scholar and teacher, she introduces students to a particular facet of climate change that they might not have learned otherwise.[21] Interviewees used this storyline of knowledge provider, along with knowledge creator and climate change leader, in ways that identify women as authorities on climate change.

EXPERTISE CALLED INTO QUESTION

However, participants also indicated several instances in which their expertise was challenged or underappreciated. For instance, Linda, who draws on her biology degree for her advocacy work on children's health in the United States, found it frustrating when people assumed that she did not have the background to speak authoritatively on the issues she was discussing. She said she feels that she needs to convince the mostly male politicians she interacts with—"cranky men," in her words—that she understands the science behind the children's health issues she addresses. She quickly pointed out that she should be treated with respect even without a science background, but that it bothered her when her credentials were not valued or acknowledged and she was treated with a dismissive "Okay, sweetheart." Constance reflected on the difficulties of not only being a "woman in science," but a woman who studies a scientific topic that has unfortunately become a political lighting rod. She mentioned that "it feels like you have to prove yourself," particularly when talking to people outside of scientific circles. For Linda and Constance, their scientific credentials do not necessarily enable them to be taken seriously when they discuss climate change with people who lack the same background. In Linda's case, she saw herself trying to overcome the same gendered trope of uninformed housewife used to belittle environmental activists such as Lois Gibbs and others (Blum 2008; Seager 1996).

Allison similarly noted a tendency for people to call her academic expertise into question:

> I do actually see, like I literally have seen the faces of people who have read my book—when they find out who I am, and that I'm female, and that I look the way I do. I look younger than I am, too. You can see that they're like, really? You're the author? And then the people who haven't read my work, and don't know who I am, they can be really difficult to sort of get past what their preconceived notions of what an expert looks like and I get like a lot of mansplaining still. My chiropractor was mansplaining to me about climate change adaptation the other day. I was like, dude. Seriously? I teach entire courses about this.

Allison was by no means the only participant to recount this kind of exchange or pattern of behavior. Janice, for example, had been working on a clean energy initiative for her US-based climate change organization for over five years when her male colleague, who was the originator of the initiative, left the organization. She was introduced to everyone on the team as the new leader on the project, and she did significant outreach with those involved on where everything stood. She indicated her extensive knowledge and work on the project to me. "I wrote the proposal for the work to be funded," she said. "I delivered the presentation to the legislature, and the academicians for part of this biofuels working group, and I wrote the follow up email." Despite all this, one of the men in the initiative sent an email to her male colleague asking questions about the work instead of her. "I often feel like I am invisible here" was her reaction. When I asked a follow-up question about who she is invisible to, she responded, "It's with men, women do not see me as invisible. Women see me as a leader with the movement."

Aida, an academic in the United States, did not claim to have experienced rejection of her specific expertise as a climate scientist, but her first reaction to the phrase "gender and climate change" was to wonder about gendered expectations of expertise by saying "people are more, maybe, trained to recognize a male expert."[22] These examples of women feeling like their expertise was not recognized or that the knowledge and contribution of other women were not recognized exhibit the women-as-knowledgeable discourse. Interviewees thought that there was specific expertise possessed by women that was being ignored, downplayed, or belittled in gendered patterns.

Women's scientific knowledge about environmental topics is related to gendered societal trends in science, technology, engineering, and math (STEM) fields. Existing research indicates that while there are no gendered differences in early quantitative and mathematical ability (Kersey et al. 2018; Kersey, Csumitta, and Cantlon 2019), there are gendered differences in science and math education, which foster gender imbalance in STEM fields (Jacobs and Simpkins 2006; VanLeuvan 2004). Girls tend to express less interest in science and math, have less confidence in their science and math abilities, and have lower expectations for success in these courses than do boys, largely independent of their actual performance in their science and math education. This is largely because the way they are introduced to these subjects reinforces gendered assumptions about which topics are appropriate for boys and which for girls (McCright 2010). At the same time, women and men tend to express different perceptions of science in the general public, with women often expressing less confidence in science overall (Fox and Firebaugh 1992).

One interviewee, Sharon, mentioned the impact of how people are educated into STEM fields:

> I've heard people . . . higher ups at organizations that are pro–clean energy, pro–climate change [action] talk about the difficulty of talking about solar as a technical issue to women who aren't as "technically minded." And those kinds of comments. And even the science you know, "There's so much science in climate change and do women really understand it because stereotypically women are not as science oriented." [Prompt: do you agree with that?] No, I think that's a product of the education system. I don't think that's because our brains are pitiful. [Laughs] We can see the push to try to get more girls into science, but this hasn't always been the case. I think [the lack of women in scientific fields] is just a product of the way society runs education and life in general.

Sharon's response speaks to why there may be more men than women in certain climate-related fields, as well as to the prevalence of the perception of women as not "scientifically minded" is across multiple sectors. She rejected this assumption by emphasizing that the problem lies with how people are educated and not with who is "scientifically minded" or not.

Seventeen interviewees mentioned that women are underrepresented in science or that they are less visible or recognized in their scientific fields. Participants observed that women have a great deal of climate change knowledge, but that their STEM fields present some challenges to having this knowledge showcased and utilized.[23] Lydia, a nonprofit worker in the United States, for instance, brought up organizing a workshop focused on women in STEM. She found that "one of the surprising things that came out of that workshop is that the women we were speaking to, who were mostly young women, we asked them what they think of a scientist and what is it that they picture. Most of them were thinking male, white lab coat . . . so nothing indicating women." For her, this indicates the need to interrogate perceptions of expertise in climate change. She went on to say that while it is essential to establish scientific consensus about the nature and effects of climate change, when a strong connection is made between climate change and science, there might be gendered impacts to assessment of where expertise lies. This is "because there's something psychologically going on that women don't innately see strong female presences in the sciences." Phillis, a climate scientist working in the UK, related this particularly to the egos of some male climate scientists:

> I've experienced a lot of scientific egos in the field as well. . . . It's the personalities, isn't it? It's like "Oh, we're a world leader in this, and we have to do these roles that are more than most scientists." . . . I'm thinking about four or five characters that I know who think, you know, "We're the most important people in the department."

Such individuals preserve their privileged positions at the expense of others and are often able to play this gatekeeping role because of gendered expectations about expertise.

Other participants added that this expectation about who speaks in science influences how they are personally viewed as scientists. Brittany has a degree in environmental science and works for an environmental nonprofit in the UK. Noting that her studies in science were male dominated, including the make-up of her peer groups, she said, "If I'm talking amongst my peers, I've got the same degree. . . . I've got the same grades. I will get less of a space to talk than the rest of them," meaning men in her position. Constance, an atmospheric scientist in the United States, similarly said that she feels underestimated: In her words, "the biggest challenge for me both personally and professionally is just not being taken seriously as a scientist. And this goes in regard to being a woman in science and also being a woman studying climate change—which is a highly debated topic these days. So, it feels like you have to prove yourself." Jasmine brought up both gender and race in her responses.

> As a woman—not only a woman but as a woman of color—I can't ignore that intersectionality when I speak about this. Fortunately, I've seen over the years, I've seen at least in academia the graduate student population at my school in environmental science [has] been pretty good. It's like 50–50. In some cases, some cohorts come in with more women than men, which is fascinating. Unfortunately, on the other end there are not really many women of color. I'm the only black woman in my cohort in the graduate sciences and environmental sciences. And it's unfortunate that the diversity isn't there in regard to that. Because there so many different perspectives we could bring to the conversation. When we see the transition from graduate students to faculty . . . that's when we see the number [decline], especially tenured professors working in science, especially climate science. I could probably count a handful of women who have a seat at the table, so to speak, in regard to their weight in contributing to climate change discussion.

These comments highlight the importance of different perspectives being part of climate change conversations as well as the specific problem of women's absence. In Jasmine's view, the fact that women of color in particular are underrepresented in her field means that useful perspectives "that other groups, specifically white men, won't even think about" are lacking. She also noted that "in the natural sciences we're not primed to think about how gender fits in with climate change. In these hard sciences we're taught the technical aspects of it. We talk about, perhaps policy, but not in that angle." This point was echoed by Lydia, who works for a solar energy organization in the United States. She left the natural sciences because she wanted a more community-focused view of environmental change:

> Initially my focus was on extreme weather, specifically with hurricanes, and I was working on hurricane modeling . . . at [an atmospheric research center]. . . . Then there was this one day that I was just modeling a huge storm in the future that was wiping out the Caribbean islands and I just . . . I don't know. Something just clicked and I wanted to make a big switch from being the scientist that was in the room modeling things to actually being on the ground and seeing things from a community perspective. And thinking about how is it that this future storm that I'm modeling, how is that actually affecting people in communities, on a more societal level. So, I got more interested in the societal aspect of things. It wasn't anything that I had an actual academic background on so I kind of took that on for myself for my Masters and decided to do a more interdisciplinary route and combine my love of weather and environmental issues and try to see how that can be fused with societal issues.

Like Jasmine, Lydia suggested that there is something about the natural sciences themselves that shapes views of climate change. Of course, this happens in other fields, but the point these participants were making was that their disciplinary lenses did not readily incorporate social aspects of climate change, including how it intersects with gender. They both regarded this as unfortunate.

Interviewees also highlighted a few major obstacles facing women in STEM fields when they try to balance work and family obligations. This illustrates connections between the women-as-knowledgeable and women-as-caregivers discourses. Jasmine brought up that women might be underrepresented in STEM because "being a woman who wants a family but also wants to conduct science—especially climate science or environmental science where you often go into the field for long period of time or away from your home for long periods of time—that's a challenge." Likewise, some participants discussed their dual roles as both scientists and mothers during our conversations. Like Jasmine, they mentioned the difficulties of balancing a demanding job with recovery after childbirth, care duties for sick children, and other facets of motherhood.

These concerns and experiences appear to be relatively widespread. In fact, a 2019 study found that parents are more likely to leave STEM jobs than their childless peers (Cech and Blair-Loy 2019). According to the same study, 43 percent of first-time mothers end up leaving their full-time position in STEM fields after the birth of their child, while 23 percent of new fathers end up leaving their position. And these findings hold irrespective of variation by discipline, race, and other demographic factors. The study attributes some of this to the fact that many scientific fields are felt to be unconducive to the realities of parenting. Given that women still do a disproportionate share of care work in many countries, it is perhaps unsurprising that they end up leaving science and math positions at greater rates than men after becoming parents. Women associated with the 500 Women Scientists group wrote a 2019 blog post about breastfeeding challenges in science, which raises some similar points. According to the authors, there is a lack of institutional support and cultural acceptance of breastfeeding in academic settings, which leads to workplace protections for breastfeeding not being honored. They singled out the natural sciences, saying that "there are many unique consequences of being in scientific fields that make it difficult to breastfeed, including: remote field sites, open laboratory spaces, working with potentially hazardous materials, unpredictable hours for experiments, teaching obligations, and many more" (McCullagh et al. 2019). Unless some of these obstacles are addressed, women may continue to be underrepresented in these fields.

Other participants noted that while they sometimes struggle with being seen as experts in science, they also must confront a lack of public support for climate change science in general or a public failure to value the expertise of the scientific community. For Constance, an atmospheric scientist in the United States, "the political attention around [climate science] has really made it challenging to get up and go around to campus and doing my research. It's almost like why bother with this research when I live in a country that doesn't believe the science?" Reflecting on the women-as-knowledgeable discourse requires thinking through not only when and how women's climate change expertise is identified, but also how this proficiency is viewed inside and outside of climate change spaces. Who is recognized as possessing climate change knowledge and what does this position afford them?

IMPLICATIONS OF THE WOMEN-AS-KNOWLEDGEABLE DISCOURSE

Within the women-as-knowledgeable discourse, climate change is both a facet of our world and a site of disagreement within society (i.e., skeptics or those who understand the facts). Climate change is also an academic or research specialty. People go to conferences on climate change, belong to academic networks on climate change, teach classes on climate change, and publish research on climate change. Finally, climate change is a policy problem that requires information as well as diplomacy to address. Climate change experts are needed to help get to effective solutions.

Where are the women? They are in climate change negotiations and policy meetings. They are in science labs, classrooms, and conferences. They are in spaces that have often been dominated by men. Participants stressed that women make valuable contributions to climate change knowledge, even though they occasionally feel that these contributions lack sufficient recognition. At the same time, the women-as-knowledgeable discourse also portrayed women in spaces that are in line with dominant gender norms—for example, in fields planting crops and in forests gathering resources. In these depictions, women gained important environmental knowledge because of their role as laborers or through domestic care work.

The overwhelmingly positive storylines associated with the women-as-knowledgeable discourse indicate a level of pride in women's contributions to climate change understanding and action. In many cases, participants used the discourse when mentioning their own climate change knowledge and activities. They outlined their projects and tasks, speaking engagements, research topics, classes taught, among other activities. They also used the discourse to refer to women more generally who have a high profile in climate spaces, largely within either diplomatic or scientific circles. Christiana Figueres in particular represents a visible example of a woman who is acknowledged as a climate change expert, by both interviewees as well as many people across the world. Although she is from the global South, she is typically discussed in terms that highlight her agency. Figueres's comes from a family of political figures from Costa Rica, with her father and brother both serving as the country's president, and her mother serving as a member of the legislative assembly as well as an ambassador to Israel. She was educated at institutions in the global North and fails to fit the profile of the rural resource user that is typically used to refer to women from the global South. Overall, women from the global South were depicted as knowledgeable because they perform specific tasks that afford them knowledge. When participants gave examples of women who are knowledgeable, it was because they are scientists, or knowledge brokers, or diplomats, and these much more often came from the global North. When we ask, "Who are the women?" there is a noticeable dividing line between the knowledgeable women of the North and South. While there were exceptions to this trend, such as participants mentioning individuals from the global South they had worked with or seen present at a conference or event, the larger trend is one of expertise, at least traditional notions of expertise, being in the North.

There is a fine line feminist scholarship typically walks in situations like these between avoiding essentialization on the one hand and identifying patterns of behavior on the other. While it is crucial that scholars and practitioners adopt a broad understanding of knowledge and expertise that incorporates local knowledge gained from gendered household tasks or resource use, it is also problematic to assume that women of the global South are knowledgeable about climate change predominantly or only because they perform these roles. One problematic effect is their knowledge can be exploited without them being incorporated as full partners in climate action (i.e., their "epistemic objectification") (Dey, Singh, and Gupta 2018; Dove 2006; Federici 2009; Sapra 2009; Tuvel 2015). Another is that it renders invisible all of the female scientists, scholars, and politicians from the global South who doubly fail to fit dominant assumptions about where expertise lies. In this way, the women-as-knowledgeable discourse could serve as a frame to counter the frequently simplistic portrayals of women as victims of climate change, but only if it is used to demonstrate the many forms of climate-relevant knowledge that exist.

According to interviewees, women's climate knowledge takes many forms. For some, being knowledgeable means realizing the scope and scale of climate change and being willing to act. For others, it means producing and sharing data about climate change with peers, policymakers, and the public. The women-as-knowledgeable discourse features storylines that highlight women as being less susceptible to climate skepticism and as contributing to global climate change research. It is a discourse marked by positive depictions of women and their qualifications to speak with authority about climate change. As such, it presents opportunities as well as obstacles to framing climate change in ways that are beneficial to the goals of sustainability and justice.

One opportunity afforded by the discourse is its broad conceptualization of knowledge. The discourse encompasses both a status quo orientation toward knowledge as well as a transformational direction. On one hand, the discourse is often associated with scientific, scholarly, or professional forms of proficiency. Participants mentioned specialized knowledge or savvy, as well as expertise indicated by credentials and title. These include chief climate scientist, head negotiator, and PhD. In this way, the discourse potentially reinforces existing ways of framing and understanding climate change. One facet of the discourse asserts that women can achieve proficiency in fields such as science, academia more broadly, and politics, all of which have historically been male-dominant. This defiant function of the discourse suggests that women need to have their contributions recognized. It does not necessarily challenge the knowledge itself or the practices through which it is gained and disseminated. It does not assess how some perspectives are regarded as "expertise" and others are not. For instance, those who reacted negatively to the idea of climate science being masculine tended to object to the notion that some might regard women as unable or unfit to participate in science. They suggested that women's knowledge and participation undercut any portrayal of these being masculine spaces. For many, their first response was to support the institution, even if they then pointed out gender imbalance in power relations or other areas. In this vein, presenting women as knowledgeable about climate science or climate politics paints them as useful to mainstream climate action as they produce knowledge and perform functions that are necessary to climate understanding and policy.

On the other hand, the women-as-knowledgeable discourse also considers women's climate knowledge as unique and potentially transformational. A few interviewees noted the "powerful" role that women play when they educate their families and communities about climate change. This task will not likely be put into a resume or be evaluated for tenure and promotion, but it still represents a form of knowledge dissemination in its own right. At the same time, while some participants identified men as being more heavily involved in technological approaches to addressing climate change, they often followed up by noting that women frequently possess the capacity to get people to care about climate change on a personal level. Interviewees mentioned that approaching climate change as a matter of rethinking the human-nature relationship—as opposed to a technical matter—is stereotypically feminine.

For example, Molly is a US-based academic who contrasted "masculine" technological approaches with a social justice frame. One reason she argued the latter is less prominent is that it is more difficult to incorporate into policymaking. "When we frame climate change as a justice issue or as an economic issue then we are getting into much more, much broader frames . . . and it becomes a, you know, even more systemic kind of way of thinking about it." For Molly, the fact that humanity is tackling climate change through existing structures influences how we think about it. Furthermore, it shapes assumptions about knowledge and expertise. If climate change is approached in a narrow fashion, then this simultaneously limits whose expertise will be consulted. Even when women's expertise or skills are not acknowledged, it does not mean they are not there. They simply may be less visible or less valued. When Candice argued that she is not rewarded for her knowledge broker role by her scientific colleagues, she was pointing out a potential limitation to existing beliefs about what kind of expertise society should value. She indicates that her ability to make climate science relevant and meaningful to the public (i.e., a social role) is also essential. Feminist scholars have likewise called for rethinking climate knowledge and expertise. According to Christina Shaheen Moosa and Nancy Tuana (2014, 678–679), "Our knowledge and ignorance concerning climate change are shaped by relations of power." Dominant ways of discussing climate change as a scientific or economic issue typically frame climate change knowledge as objective and value-free. Despite these representations, "the methodologies and rhetorics of climate science, like all sciences, are far from value-free. The solution, of course, is not to remove values from science, but to render them transparent and subject them to careful analysis." Likewise, feminist economists such as Julie Nelson (2007) claim that economic frames of climate change tend to overlook those who are most vulnerable to climate change as well as future others. This scholarship challenges the idea that scholars and practitioners should add more women or highlight women's contributions to existing ways of knowing climate change and instead critically engage with those ways of knowing.

These points about critical evaluations of knowledge relates to interviewees reflecting on proficiency more broadly. In some cases, expertise was depicted as a necessary qualification for entry into a group, space, or circle of conversations. Jasmine justified her decision to become a scientist "to show my expertise in it, so the weight of what I say is valued." This indicated her assumption, shared by many others, that her perspective will be less respected if she lacks a specific set of credentials. The value of her expertise is evaluated externally. Other participants reflected on internal evaluations of their skill or proficiency. For instance, Alice has had experience as a practitioner with the Scottish government as well as with an organization that provides research and analysis to the government on climate change for policy purposes. She has been in her position for about two and a half years and explained that she has had an interest in climate change for fifteen to twenty years. Despite all this, she claimed that

> My personal challenge is that I always feel that I need to know more before I can do things properly. And I never know enough partly because I can be dealing with [one issue] in the morning and in the afternoon go on to [a very different issue]. I just never know enough. I don't know what that says about gender but that's my personal challenge.

In Alice's case, she perceived her own expertise to be insufficient because she is responsible for knowing something about a diverse range of topics related to climate change. This reflection points to the complexity of climate change and the enormity of the number of topics it intersects with. Being expected to know "enough" about even a single facet of climate change can be very daunting.

So, where does expertise come from? How are we trained or socialized to think about knowledge and who possesses it? A 2016 study of youth climate activists in Canada by Joe Curnow and Jody R. Chan (2016) argues that gender plays a central role in how expertise is perceived in activist spaces, with men who behaved in "typically masculine ways" being more likely to have their positions praised and reinforced and being recognized as a leader or expert. They claim that "becoming an expert in a community of practice may have less to do with adopting shared practices or acquiring new skills, and more to do with performing masculinity, even in spaces that value social justice and anti-oppression" (Curnow and Chan 2016, 146). In other words, just as gender influences perceptions of knowledge among older generations, similar patterns may be at play in the vibrant youth climate change movements across the globe.

Asking about where expertise comes from is obviously relevant for thinking through expertise about climate change, but it also speaks to patterns I found of interviewees worrying about their capacity to answer my specific questions. There were multiple instances of participants stating that they were not experts on gender and climate change. They told me that their responses were "just their opinions," that "they had not done any research on this," or their "understanding of that is very, very limited at this point." Some apologized for their answers,[24] saying "sorry" for not being able to recall specific pieces of information, or for going off on tangents or "waffling," despite the fact that they were just answering the questions that I asked.[25]

Several interviewees also seemed hesitant to make strong claims because they felt that they did not have the background or credentials to back up their claims. I even had one participant tell me that there were other women in her nonprofit organization who were interested in my project and would have liked to participate, but that they did not feel comfortable doing so because they did not have a background in topics like gender and climate change. While those scholars who have given us decades of important gender and environment scholarship obviously have a great deal to tell us about gender and climate change, I argue that my participants working on climate change issues also have some important insight. It would not do to replace one form of knowledge with another, but the data from the interviews allow us to understand how a range of people who identify as women make sense of these connections based on their work and everyday experiences. It is unfortunate that others were turned off from participating because they negatively evaluated their own ability to make a contribution to my project. I wonder what the larger implications of these negative evaluations of expertise might be for other areas of climate action. Is the behavior of these interviewees just part of a larger pattern of women feeling uncomfortable stepping out of an area in which they have specific credentials or expertise? Do scholars and practitioners limit themselves if we assess expertise so strictly?

These questions are particularly relevant when considering the close association between climate change and scientific knowledge, as argued by multiple participants as well as scholars. According to Karin Bäckstrand (2004, 695), "Environmental politics is increasingly scientised and technocratic. The scientisation of environmental politics implies that political and social issues are better resolved by technical expert systems than democratic deliberation." If our understanding of a climate change expert is of a "scientist," then there are important gendered ramifications of this, not to mention raced, classed, and placed implications (Bäckstrand 2003, 2004; Israel and Sachs 2012; Moosa and Tuana 2014; Nagel 2016; Tuana 2013). For instance, interviewees noted that institutions such as the Intergovernmental Panel on Climate Change (IPCC) are male dominated. They mentioned the IPCC specifically as an entity that has a high degree of prestige in climate change discussions, but one that has a significant gender imbalance.[26] A 2018 study of gender and the IPCC found that women not only confront challenges based on gender, but that they also "face multiple and diverse barriers associated with social identifiers such as race, nationality, command of English, and disciplinary affiliation" (Gay-Antaki and Liverman 2018, 2060). A study by Miriam Gay-Antaki and Diana Liverman draws on survey data from women who have participated in authoring IPCC reports and thus shines light on first-hand accounts of obstacles some women encounter in having their expertise valued. It finds that certain categories of women struggle particularly hard to feel heard, including women with young children, women of color, and women from the global South. Taking gender seriously in discussions of climate change entails interrogating where current systems might perpetuate marginalization. For some women, this includes high-profile institutions such as the IPCC (Yeo 2018).[27] The IPCC is an example of an organization that is globally recognized for its contributions to climate change understanding and has taken steps to address inclusivity and diversity. However, it is also an organization that has faced criticisms for replicating existing patterns of sourcing knowledge, thus also replicating gendered, raced, and classed patterns of power in climate spaces (Nhamo and Nhamo 2017).

None of this is to deny the value of the crucial work done by climate scientists, but rather to think through ways in which gender plays a role in legitimizing some voices more than others. Some climate scientists have reported intense gendered or sexualized harassment when their research gets picked up by mainstream media or climate skeptic outlets. While most of the harassment takes the form of social media posts, emails, and phone calls, some worry for their physical safety (Johnson, Widnall, and Benya 2018; Ogburn 2014; Waldman and Heikkinen 2018). Although none of my participants described these extreme levels of harassment, they did remark on belittling YouTube comments on their lectures and other disparaging behavior.

The "masculinization of environmentalism" that Sherilyn MacGregor (2009) critiques results in male-dominated scientific fields mostly closely associated with climate change expertise. The ramifications of this might be gendered harassment of female scientists who fail to fit the dominant profile of one who has scientific expertise. This notion of climate expertise can reinforce existing models of knowing and acting without leaving room to reflect on possible transformations that could render climate action more effective and just. This is just what interviewees suggested when they highlighted the transformative potential of women's climate change knowledge. None of the participants who used the transformational storylines suggested that climate science was not valuable. Rather, they stressed that failing to connect climate science with climate change impacts in people's daily lives would leave climate action either ineffective, unjust, or both.

6 WOMEN AS AGENTS: PARTICIPATION AND BARRIERS IN CLIMATE CHANGE WORK

Abigail was one of the first participants I spoke to. When I asked her how long she had been doing environmental or climate change work, she recounted a professional history that saw several twists and turns. She went to New Orleans, Louisiana, to help with the extended recovery efforts a year and a half after Hurricane Katrina.[1] She intended to stay for two weeks on a January term trip from school, but instead stayed for over a year. She eventually left to participate in Green Corps, a training program for recent college graduates to learn how to run environmental campaigns. For a year, she worked all over the United States on clean energy campaigns. Her next position involved working to start a new project to diversify the coal economy in rural West Virginia through renewable energy. She then moved to Atlanta, Georgia, to work on stopping new coal plants from being built and began working with her current environmental organization. Abigail is an example of an interviewee who has been very active in a variety of roles in environmental and climate change work. At the same time, she saw numerous ways in which women's involvement in this work can be blocked. In her words:

> In terms of structures, decision-makers are still overwhelmingly male. For example, the public service commission is five white men over the age of fifty. . . . So, I think women are largely underrepresented in the power structure that we're trying to affect. That's probably, I imagine, also true in the utility industry. And then I think about the dynamics of the nonprofit sector in this country. . . . So, I see a lot of the work, like in the trenches, being done by women. Although [I] also recognize in the nonprofit sector, leadership often skews male.

Abigail highlighted various structures that influence whether and how women are likely to be active in climate action. She also pointed to gendered patterns in the decision-making that influences climate change.

Abigail's work history fits in with the *women-as-agents discourse.* When using this discourse, participants spoke with pride about women's involvement in multiple facets of climate work.[2] They identified a range of specific roles and jobs, including women as consumers, laborers, wage earners, nonprofit workers, engineers, social scientists, climate change negotiators, farmers, conservation workers, teachers (of children, in their communities), and decision-makers, among others. They also identified women as potential agents in fields such as politics, engineering, and science—as space in traditionally male-dominated fields opens up for them. Women were frequently described as active participants in either environmental harm or action to protect the environment. The women-as-agents discourse also focuses on ways that women face obstacles to participation or feel sidelined or disregarded when engaging in climate change action. I highlight multiple storylines related to the idea of women being constrained, limited, or blocked, ranging from women's literal absence in some climate change spaces to women feeling that their contributions are undervalued. The discourse is consistent with calls to critically reflect on women as multifaceted environmental agents. It counters the portrayal of women as something that environmental change happens to and instead illustrates participation and action.

AGENCY AND CLIMATE CHANGE

Academic fields such as international relations have historically tended to privilege states as agents within the international system (Ford 2003). Global environmental crises that do not respect state boundaries, including climate change, illustrate the necessity of thinking of agency well beyond states. Many scholars have pondered the concept of agency in international relations and environmental politics, considering questions such as which actors have agency (Ford 2003; Maclean 1999), what blocks agency (Brown and Westaway 2011), and how agents and structures interact (O'Neill, Balsiger, and VanDeveer 2004). One debate has centered on whether and how we should think of agency at the individual level. Some scholars express skepticism about or caution over focusing on individuals as agents to address environmental change (Kent 2009; Maniates 2001). Michael Maniates (2001, 33) famously warned against the "individualization of responsibility," arguing that "when responsibility for environmental problems is individualized, there is little room to ponder institutions, the nature and exercise of political power, or ways of collectively changing the distribution of power and influence in society—to, in other words, 'think institutionally.'" He and others warn that focusing on individuals could result in the depoliticization of environmental issues.

Over time, however, global environmental politics scholarship has focused more on the individual and the "everyday" in environmental action. Scholars in this vein present an optimistic view of the transformational potential of individual action or at least a recognition of the everyday as a site of environmental politics (Eckersley 2020; O'Brien 2015). Karen O'Brien (2015, 1170), for instance, argues that "because of the scope and scale of transformations needed to reduce climate change risks and vulnerability, a more expansive view of political agency is required—a view that captures an individual's ability to contribute to transformations both by changing behavior and by influencing structures and systems." This view acknowledges individuals as political actors, but often fails to conceptualize them as gendered, raced, classed, and otherwise situated within sites of power (MacGregor 2021). According to Sherilyn MacGregor (2021, 48), "seen through a feminist lens, it appears odd that academics would choose to focus on everyday practices—especially those that are empirically most likely to be unpaid and feminized—without asking questions about who is performing them, under what conditions, and to what extent at their own choosing."

Feminist scholars across multiple disciplines have long centered the everyday in understanding gendered power relations within and across the structures that make up the international system (Di Chiro 2008; MacGregor 2017b). For instance, research has examined gendered environmental behavior such as green consumption as a site of agency in the environmental realm (Kennedy and Kmec 2018; Stolle and Micheletti 2005). This work acknowledges the empirical finding that women tend to outperform men in pro-environmental behavior (Hunter, Hatch, and Johnson 2004) but goes beyond this to call on scholars to take gender into account when we theorize and research environmental agency. Dietlind Stolle and Michele Micheletti (2005, 45) argue that political consumerism, or "the buying or boycotting of products and services based on political or ethical values" is a form of civic engagement and political participation that is disproportionately undertaken by women and should receive greater attention in attempts to understand participation and social interaction.

This work is consistent with feminist scholarship that complicates the idea of "women's agency" more broadly. Feminist scholars who work on violence have demonstrated that women's agency is complex and often violates dominant assumptions about women's acceptable behavior. While gender norms in many societies paint women as peaceful or vulnerable, in reality, women willingly and often enthusiastically engage in political violence. Dominant ways of framing agency often ignore or mischaracterize these violent women (Åhäll 2012; Gentry and Sjoberg 2015).[3] Likewise, feminist environmental scholars show that although women's agency takes multiple forms in climate change action, essentialized portrayals of women and climate change that limit understanding of women's action as either virtuous or based on their victimhood persist (Arora-Jonsson 2011). Lena Partzsch (2017, 197) conceptualizes one form of agency as "power to" or the ability of individuals or separate groups to get things done.[4] Some of these choices are conscious efforts to mitigate or adapt to climate change, while others are choices that are freely made, but result in environmental harm. Agency in the realm of climate change, then, involves making choices and taking various actions that can either positively or negatively impact efforts to halt climate change.

WOMEN'S PARTICIPATION IN CLIMATE CHANGE ADVOCACY

Media stories, government and other reports, and policymaker speeches frequently emphasize the importance of female participation and leadership in climate change and environmental work. For instance, in a series of public speeches in 2010, UN Secretary-General Ban Ki-moon repeatedly remarked that "the world's women are the key to sustainable development, peace and security." Additionally, articles with titles such as "46 Sustainability Leaders (Who Are Also Women)" (Townsend 2018), "8 Badass Environmentalists You Should Know" (Pierrat 2015), "The 20 Most Influential Women in Green" (Rogers 2010), and "4 Black Women Leaders on Climate, Justice, and the Green 'Promised Land'" (Calma and Rosa-Aquino 2019) highlight the contributions that women are making to environmental and climate change action. These examples illustrate that women's influential role in climate change is well-known and often celebrated. My interview participants were no less excited by the considerable presence of women in environmental activism and advocacy. A central storyline in the women-as-agents discourse is women-as-active-participants in environmental organizations and the environmental movement more broadly.

Large numbers of women are engaged in environmental action across the international community and have played this role for decades if not centuries in many countries. In the United States, for instance, where many of the interviewees live, women have advocated for environmental protection in large numbers since the turn of the twentieth century. They have been particularly active in advocating for environmental justice through the anti-toxics and anti-environmental racism movements since the 1970s (Bullard 2005; Seager 1996; Stover and Cable 2017). Given that many of the interview participants work for environmental nonprofit organizations, it is unsurprising that women's participation in climate change advocacy and nonprofit organizations came up a great deal in the course of our conversations. Interviewees often reflected on their personal experiences with environmental or climate change advocacy. Brittany, for one, works for a climate change organization in London that has a large number of women on its staff. She remarked that this is very encouraging for her as a female employee. In her words, "I think personally about my own realm, how it changed, but also about a bunch of climate bad assess that there are out there . . . loads of women working in that space. And I guess it's a lot more gender equal than a lot of other sectors." Likewise, Meg and Haley commented on the large number of women in their respective environmental organizations in the United States. Meg, who works in an environmental conservation organization, said she finds it easier to problem-solve in an environment that has gender balance. Haley, who works for a solar energy nonprofit, noticed a shift in the gender balance around her when she moved from working in the solar industry to working on solar advocacy, with women being more heavily represented on the advocacy side. When participants noted the abundance of women working on environmental issues, this was typically said with pride. They were proud to be in a field where women are particularly active.

While many interviewees highlighted women's overall involvement,[5] several also pointed out that there are gendered patterns with regard to where women are most active in climate advocacy. Haley specifically mentioned organizing, marketing, communication, and outreach as the areas in which women are present in large numbers in her organization and noted that they most frequently occupy middle-level positions. Alice, who works for a climate nonprofit in the UK, also noticed women in more middle- and lower-level positions. She said that her team is all female, but the management board is all male, as her organization has "become aware." Heather, a US academic, pointed out where women's voices seem to be the loudest:

> I'm just thinking of social activism and there are a lot more women's voices. . . . Not the big global stage where you have the Michael Manns and stuff, but, when I think about indigenous communities speaking out about climate change, women are visible in those communities. . . . There's so few women leaders internationally at every level, so yeah, it seems like if women are influencing . . . it's coming from sort of the bottom up.

Rebecca, a nonprofit worker in the UK, echoed several of these points, saying "my area has been very much in the communications and behavior change arena, and we do typically find more females work in that area than the other side of things which might be engineers or more technical stuff which does become more male dominated." In effect, then, women are often absent from those areas related to scientific expertise, as examined in chapter 5. Rebecca also reinforced the argument that women occupy lower- and middle-level positions more often than senior positions at many nonprofits. She went on to attribute some of this to some women getting burned out when working in that sector:

> And something that I feel quite passionate about because I was one of them, I went into a charity very eager to please, very passionate about what I believed, and kind of loved what I did but was completely overworked and underpaid. Not that that would be any different whether I was a man or a woman, but I feel like that's quite a female trait maybe, . . . to not command some space in a "I deserve it" kind of way. . . . I was one of them and I feel like I see a lot of charities and a lot of organizations doing that to young people. I guess to men and women, but I see it more predominantly in women. They are eager to please, become completely overworked, they burn out and crash. Then they move on to a different sector or have a massive break and find their way back in.

Rebecca regarded people's excitement and eagerness as a good thing for getting people involved in environmental work, but also something that potentially contributes to some getting overworked and potentially even leaving the field. In her view, this often happens to young people, particularly young women.

Participants listed multiple reasons for their own decision to get involved in climate change or environmental advocacy. This was the case for those who worked in the nonprofit sector, as well as those who played a less formal advocacy role. Echoing Rebecca's statements above, most participants who work for nonprofits said that they wanted a job in which they "made a difference." Participants also often identified a long history of environmental awareness or activity that eventually led to them seeking a job in an environmental organization. Abigail, a US-based nonprofit worker, identified herself as an activist from the early age of eleven, when she started an environmental organization comprised of herself and her best friend called "Environment Crime Zappers." She explained that she "was already predisposed" to environmental action:

> I grew up thinking that every kid was plunked on the back of a bicycle and everyone had a compost pile in their backyard. And I grew up reading *Ranger Rick Magazine*, . . . [and] I attribute a lot of this whole orientation to the world as coming from that early exposure. . . . It turns out that my dad in his early twenties was out in New Mexico helping start the first solar association down there as an AmeriCorps volunteer. My mom was an educator; she worked in Head Start early on in her career. So, I think it's pretty deep in my family to have some kind of connection to people and the world around us.

Other participants also noted their upbringing as a factor in their environmentalism. Shea dates her interest in environmental and climate change issues back to her time in college, when "a history professor mentioned the term 'environmental justice' to me and I thought that term was pretty cool." After looking it up, she "really got hooked to the notion of environmental justice and climate change policy." This led her becoming an environmental attorney and working for a consumer-focused clean energy organization in the United States. Shea was among several participants who identified school as the place that inspired their environmental interest. These women recounted classes they took, clubs they joined, or teachers they interacted with that fostered their environmental awareness. For some of these participants, this was the start of multiple years if not decades of environmental activism or advocacy.

Brooke, on the other hand, is an example of an academic who told me that she had recently "become an activist." She attended a city council meeting and spoke in favor of her city signing on to a national US mayor's letter on climate change. In explaining her idea of activism, she also mentioned her role spearheading a faculty climate statement at her university. Sarah is another US-based academic who recently played an activist role in her community. She spoke up at her daughter's school about how climate change was being portrayed there.[6] Her daughter's class was shown a pro-fracking and an anti-fracking movie, but in her view, the anti-fracking perspective was particularly weak. It bothered her that the students were receiving biased information, so she decided to take action by pushing back against what she saw as inaccurate climate information—backed up by her knowledge as a scientist.

The previous examples highlight the breadth of inspiration and action that participants identified as being associated with climate change agency. Overwhelmingly, participants argued that women's action and participation is necessary and essential for effectively addressing climate change. This necessity is born out of women's expertise and skills, but also because of the sheer fact that they are human beings who contribute to and are impacted by climate change. Participants said things like "They're half the population!" They also mentioned that women are more likely to have longer lifespans and that women outnumber men globally.[7] For these participants, women should be included and acknowledged in climate action simply by virtue of their presence and humanity.

Lyra, who works for a climate organization in Germany, noted that "if we want to tackle climate change and we do not really involve one half of the population, it's kind of a no-brainer that we really cannot achieve this paradigm shift towards a low-carbon society." She went on argue that society needed to include women in climate change projects not as simply a box to check or a grant requirement to meet, but because they have to be seen "as agents of change." Lyra was not alone in using this kind of phrase. Farhana thought back to her UK-based environmental organization's recent trip to Bangladesh when she repeatedly referred to women as "change agents," and "strong advocates within their communities." This perspective recognizes women's agency rather than viewing them as people *to bring in*. It identifies women's choices to get involved and sees them as already making important contributions to climate action.

When thinking about who takes action on climate change, participants mentioned women working at multiple levels—from women providing food for families, to women working in their communities, to women leading global climate negotiations. The scope and scale of action were huge. Participants recounted many specific examples of women in communities in the global South playing central roles in environmental management or development projects. For instance, Selma, who works for an environmental organization in the United Kingdom, recounted working with widows in Guatemala who made and sold weavings at festivals in California and used the proceeds to build water systems and latrines for their villages. In the course of a discussion about women's participation in climate change solutions, Lyra, who works in the nonprofit sector in Germany, recounted a story of a solar project in Latin America that failed to be sustainable until women were included in the project in greater numbers. Once women were trained to maintain the technology and took ownership of the project, it was more successful. She viewed this as an example of the necessity of including women in all facets of climate change work, while also noting that their potential was not fully harnessed in the process of designing climate projects or climate policies.

At an individual level, participants mentioned both women who are known for their climate action and those who are not. As discussed in chapter 5, seven participants specifically identified Christiana Figueres as a woman whose climate change action is visible and essential. Participants also mentioned previous female heads of the US Environmental Protection Agency and of European Green Parties, as well as individuals such as Wangari Maathai and Mary Robinson as examples of women playing a prominent role at the state or international level. However, some were also quick to note that there are also women at the international level who do not advocate strongly for climate change. Mary, a British student and nonprofit worker used former British Prime Minister Theresa May as an example of this. "I think whilst it is important to have gender balance in decision making, unless the women and men that are making those decisions are doing it in a gender sensitive way, it won't necessarily change anything." Similarly, Allison, a US-based academic, argued that

> In politics, right, we're kinda depending on guys because that's what's out there. If more women would run, that would be awesome. But then again, I don't know that that's necessarily a guarantee that they'll do anything differently. I mean, you can look at Theresa May right now and Margaret Thatcher in the past.

Women, in short, are complex agents rather than simplistic "environmental protectors."

CONSUMPTION AS AGENCY

When thinking of women's agency regarding climate change, there is often a tendency to assign them benevolent roles. In media stories and policymakers' speeches, women are often depicted as environmental and climate change saviors (Arora-Jonsson 2009, 2011; Awumbila and Momsen 1995; MacGregor 2017a).[8] Depictions of women as "Gaia's warriors" or environmental givers of life proliferate in both the popular imagination and at one time among scholars, too (Sandilands 1999). For instance, early versions of ecofeminism used essentialist visions of women's closeness to nature to depict their central role in environmental protection as natural. After coming under significant critique from both within and outside of gender and environment scholarship, representations of women's place in environmental debates have become much more complex (MacGregor 2017b). At the same time, environmental protection still tends to be associated with femininity by many (Swim, Gillis, and Hamaty 2020).

One storyline from my interviews that counters or at least complicates this positive representation is that of women-as-consumers. Scholars have focused a great deal of attention on consumption and its role in environmental change (Dauvergne 2008, 2010; Princen 2002). While all living things consume resources in order to live, the effects of overconsumption or misconsumption are what concerns environmental advocates and academics. Thomas Princen (2002, 33) explains that *overconsumption* is "the level or quality of consumption that undermines a species' own life-support system and for which individuals and collectivities have choices in their consuming patterns," while *misconsumption* is the practice of individuals consuming in a way that undermines their well-being "even if there are no aggregate effects on the population or species." These specific forms of consumption have been the subject of both academic and policy debate. However, to date there has been little attention paid to connections between consumption and gender in fields such as global environmental politics, although work from multiple academic fields that finds that socially conditioned roles and responsibilities do influence consumption (Casey and Martens 2007; Ghodsee 2007; Johnsson-Latham 2006; True 2003).

The women-as-consumers storyline encompasses a depiction of women making decisions and taking actions that can result in environmental harm or good. Consumption was sometimes tied to gendered divisions of labor or norms in society. For a minority of participants, it was a source of women's empowerment, whereas for most it was indicative of the sometimes negative consequences of women's incorporation into larger economic and social structures. One common manifestation of the storyline involved noting women's specific role in making purchasing and other financial decisions for families or households. For instance, Haley, a nonprofit worker based in the United States, referred to women as the "chief financial decision-makers" of households. She went on to argue that this position would likely mean that women would be aware of the effects of climate change more immediately than would someone in a different position. Likewise, Alice, a UK-based nonprofit worker, mentioned women's tendency to be responsible for purchasing food for families and how this would cause them to notice an increase in food prices linked to climate change. For Haley and Alice, the gendered divisions of labor in the household result in women having advanced knowledge or experience of specific climate change impacts—namely those connected with food and other household goods.

Annica, a US-based academic, highlighted gendered purchasing patterns by arguing that women probably have a greater say in the purchase of appliances. "Definitely the ones that are located in the household. . . . So, I see a gender division in terms of purchasing decisions that influence household level purchases." For Annica, women's consumer decision-making extended to household items, but possibly not to other purchases in the same way—like buying a car. She clarified that she thinks men just tend to care more about what kind of car they drive.

> I'm really not interested in what car we drive. I'll drive whatever we have at home. I obviously, because I'm active in the area of climate change, I am more excited if we drive a hybrid than not, but whether that's one kind of hybrid or another kind of hybrid that really doesn't matter much to me. Whereas I see, not necessarily in my family but in other families, at least in the US and also in Europe, a car purchase is more important for a man in a household. So, it's not necessarily division of labor, it's what matters to people. What they're interested in.

Although Annica identified gendered patterns in consumption, she did not interrogate where they come from. In her view, her husband, along with other men, just care more about cars then she does. She doesn't identify similar levels of interest in the appliances that women tend to have a greater say over. Men's consumption is assigned to their interest, but the motivation for women's consumption in the household goes unassigned.

These references to their own role as consumers or how consumption plays out in their households were typical across the interviews. Nicola formally worked in the nonprofit sector for an environmental organization in the United States, but also mentioned actions she takes outside of work, including being involved in the fossil fuel divestment campaign at her alma mater and being "conscious of my own impact through my dietary choices and, sort of, consumption choices." Nicola noted that these choices about consumption are "very much part of my identity and thinking about the impact in the city that I live." Not only did she consider green consumption to be an important choice for her, but she saw it as part of her identity. Lily, a UK-based nonprofit worker, claimed that "I'm trying to make a difference and have lots of silly internal struggles in the supermarket. I'm going, 'Should I buy this pepper that's not wrapped in plastic or should I buy these three peppers that are wrapped in plastic but are much cheaper?'" Selma, a nonprofit worker in Germany, said that she doesn't always have time to find the greenest options since she is a busy, single mother.

> I'm thinking about women and consumption, in the Western world much more, and the difficulty that, for example, I would have. I'm a single mom, so I'm very aware of my consumption. Not 100 percent, but 50 percent aware of what I buy and should buy, and [what] the footprint would be. But at the same time, I cannot afford the time to run around and find the right place to consume better. So, I think in terms of mitigation, women with maybe more responsibilities in terms of raising families and maybe juggling work and family, etc., might . . . be polluting more and emitting more pollution. And I often struggle with how are we supposed to do it? You know, it's something else that we need to think about, something more that we need to do and juggle and how do we do it?

In each of these examples, interviewees reflected on how their individual consumption patterns fit within their larger efforts to address climate change. They were also frequently infused with a sense of guilt about consuming. Brooke, an environmental politics scholar in the United States, voiced this outright by saying that she feels like a hypocrite because she doesn't live a carbon-neutral life. Darcy, another US academic, touched on this emotional component of consumption and trying to be "green":

> I think [women have] certainly been socialized to have a more caregiving role, I suppose, and maybe to think more about their roles as those who cook, and those who shop, and going to the grocery store, and recycling, and being environmentally friendly, and all that kind of stuff I think falls to women in the

> domestic sphere more than men. And so, it makes sense maybe that they would be confronted in a more day-to-day level with those kinds of ethical decisions. And so, I guess if they're the ones going to the store and purchasing the products more so, or the ones who would have be sorting the recycling from the garbage or whatever, on the day-to-day experiential level I think it makes sense that women would be more attuned to our environmental crisis.

Darcy's comments specifically highlighted the influence of the socially constructed "domestic sphere" in women's relationship to consumption and their feelings about the process, rather than any natural inclination to shop and cook. At the same time, these comments, along with the women-as-consumers storyline in general, depict a middle-class perception of consumption within capitalist economic structures. In this perception, women have the means and choice to make decisions about their consumption.

Allison, an environmental politics scholar in the United States, expressed frustration with the gendered nature of expectations about who in society are consumers. She referenced the "feminization of the shopaholic" and the "stereotype that women are the ones who are always buying more and more stuff." The association of women with shopping has long historical roots (Styles and Vickery 2007; Vickery 2006). Allison argued that these are likely not accurate pictures of consumption, and there is some academic evidence substantiating her point. A 2013 study of consumption patterns in the United States by Brenda Segal and Jeffrey Podoshen finds that men outscored women in both materialism and conspicuous consumption, while women were more likely to engage in impulse buying (Segal and Podoshen 2013). Their findings suggest that a simplistic picture of the shopaholic woman is inaccurate and fails to take into account men's significant levels of consumption. These findings are supported by other academic work that identifies a complex relationship between gender and consumption—not one in which women are the consumers and men are not, but rather one in which all consumption choices are influenced by multiple factors (Costa Pinto et al. 2014). In fact, consumption is a part of larger social and economic structures that we all live within. How and what we consume are deeply connected to where we live (and the economic structures there), how much money we have, and norms around things like "the good life" (Dauvergne 2005, 2010). Gender is also part of this story, as are a lot of other factors. Depicting women as somehow more predisposed to shopping than men is not only inaccurate, but also diminishes the agency of women by framing their consumption as more natural than men's.

When consumption is discussed in environmental debates, it frequently has a negative connotation. In the context of my interviews, there was often an implicit assumption that consumption is bad or part of the problem of climate change. One exception to this pattern came from a US-based non-profit worker named Gwen, who considered consumption to be a force for change, since "women have a lot of buying power." Gwen was one of only a few participants who used a consumption-as-empowering storyline. She argued, "I think that women have a role to play, you know, on policy, but they also have a role to play on changing their own circumstances; buy an electric vehicle, putting solar panels on their roof." She gave the example of a California-based solar company that markets explicitly to women as evidence that their consumption can be a force for change. She commented that "we're calling them the Mary Kay of solar panels," referencing the US-based cosmetics company that also advocates the vision of empowered women buying and selling for their own futures. While both companies have their critics, the example of a women-focused solar company fits in with an optimistic image of green consumption. Notions of green consumption, sustainable consumption, or ethical consumption encourage individuals to use goods and services in ways that enhance current quality of life while not jeopardizing the needs of future generations (Barr, Gilg, and Shaw 2011; Fuchs and Lorek 2005; Lewis and Potter 2011). Some versions of sustainable consumption are relatively status quo oriented while others see consumption as "a site of new forms of political engagement, ethical consideration and aesthetic representation" (Soper 2009, 92–93). These various forms of sustainable consumption arose out of debates about the role of capitalism in environmental change and offer illustrations of the complex relationship between agency and structure in the realm of consumption. While it is true that some individual consumers can choose to buy electric vehicles or put solar panels on their roofs, many cannot due to cost or availability. Additionally, consumers have little to no say in where solar or electric vehicle companies source or dispose of their materials (Dauvergne 2008). In the area of consumption, humans more often have the appearance of agency more than the actuality. That being said, concepts such as "political consumerism" recognize a specific form of agency that often intersects with gendered divisions of labor in households or gendered norms about consumption (MacGregor 2021; Stolle and Micheletti 2005).

WOMEN'S ABSENCE FROM CLIMATE WORK

The women-as-agents discourse features storylines that highlight women's presence and activities, but also instances where women are restricted, blocked, or absent. What does it mean to refer to women as absent from climate change spaces? It surely does not mean that women do not exist in communities that experience climate change. Rather, in the women-as-absent storyline, women are underrepresented in circles of power. This includes being excluded from debates about how climate change is conceptualized and from left out of the decisions that influence mitigation and adaptation policy.[9] Interviewees commented on women lacking "a seat at the table" or not having "their voices heard" in various spaces related to climate change, including industries that emit greenhouse gases, clean energy startups, science (or STEM fields in general), corporations, environmental organizations, policymaking groups, and conferences and negotiations. Eva, for example, mentioned these patterns when reflecting on her work for an environmental nonprofit in Germany. In her words:

> The energy world and the, especially the policy, legal world in Germany is a little bit more conservative and is mostly dominated by senior, I would say, white men. This is an issue that all my colleagues also run into, like wow this was a great conference but unfortunately, there were no female speakers, or I was the only female participant. That's unfortunately a recurring theme.

Like Eva, Talia also works in the nonprofit sector in Germany and highlighted how underrepresentation in climate events is gendered and raced. She remarked on the participation of white men at the Conference of the Parties (COP) and the United Nations Framework Convention on Climate Change (UNFCCC) and specifically referred to women's perspectives as "missing" from these meetings. This idea of visibility and acknowledgement was a recurring theme throughout the interviews. Some interviewees observed that women and other historically marginalized communities are most frequently left out of highly visible climate change work. This is one instance where some participants included an intersectional lens in their responses. They referred to gender imbalance but specifically noted that white men were the ones who tend to dominate.

Interviewees identified women in the global South as well as indigenous women as two categories who are much more likely to be excluded or underrepresented in high-profile climate work or environmental decision-making. To these, Lyra, who works for an environmental nonprofit in Germany, also added the lack of representation of "vulnerable groups in climate change processes and decision-making." Selma also works in the nonprofit sector in Germany and drew on environmental work she had done in various countries to make a similar point.

> I think it also has to do with the power of decision because in many countries, the power of decision-making is taken away from women, or at least they can't express it in public. A lot of decisions are made that are counterintuitive for women or are against what they would have done. And so, there's a situation where, in my experience, the woman sees what is going on around her. And I'm thinking a lot about East Africa and drought, and Niger, and also the villages in Guatemala, I worked a lot with widows there. They know exactly what is going on and what is needed. And it's very difficult to get it going because they're not on the circuits. They're not included in where the money is, where the discussions are, where the decisions are made.

Carin, an academic in the UK, likewise drew on her fieldwork to discuss the phenomenon of women's representation on governing councils in India. She noted a rule stating that women must make up a certain percentage of the village council. Yet they never actually attend the meetings as their participation is symbolic. Hildi, who works for a nonprofit in the UK, cited her academic fieldwork in Gambia to reflect on women's absence in decision-making:

> So, I basically do feel like women are underrepresented in the actual decision-making . . . in Gambia also, this particular village or town, it's kind of divided

into four traditional areas within the community and each of those, the oldest man in each of those will automatically sit on the local counsel. So, it's not that women are not able to access governance, but there's already a type of stepping-stone just based on your gender—you've been disqualified.

Hildi situated this example in a larger discussion of gendered patterns of access to and participation in decision-making circles.

Like Hildi, Carin, and other interviewees, identified a number of reasons for women's exclusion, most of which relate to the continuity of existing power dynamics in the international system. Participants depicted women as having less power, fewer resources, lower status in society, fewer rights, less respect, and lower literacy rates (in the global South). Interviewees argued that women have less advantageous employment (that is, they have part-time, lower paid, lower status jobs), tend to be lower income/poorer, and have historically been encouraged to take a back seat to men's success. They also pointed out existing power dynamics in climate change spaces to explain women's exclusion as white men still dominate. Participants used various phrases to refer to those who dominate conversations around climate such as "men in suits," "good old boys clubs," and "silverbacks." Haley and Sharon both mentioned "good old boys clubs" when talking about who they interact with in their work for a clean energy organization in the southern United States. In this region, the phrase refers to men in power who reward their friends and maintain the status quo, along with their own position in it. In these instances, interviewees were talking about the power structure in the utilities sector and state legislatures. Heather, who works in the United States, also pointed out the dominance in climate change meetings of "silverbacks," a term that may come from her background in biology and conservation. She recounted a meeting she had just attended that "really was old white guys in these panels":

And I realize—wow, these guys aren't even thinking about, like, it's not even on their radar to consider women's voices. What they're doing, in part . . . it's a way of, sort of, elevating their status by surrounding themselves with [other] white men in these panels. Because they're like, "These are what I think of as, really heavy hitters and so if I'm on a panel with these guys then, I too am a heavy hitter." They don't think of women as being legitimate and powerful voices. So,

> it would lower their status. I don't think they're doing this consciously at all, but subconsciously I really believe they think it would lower their status if they included women on these panels. Because then it would be women's work and it would not be as high status, or, or hardcore, and it's just infuriating.

Whether it is silverbacks or good old boys, multiple participants highlighted not only what the existing power structure looks like, but also how it is maintained by excluding those who are not currently in the dominant group. Their reflections also indicated the difficulty faced by those who seek to change those structures or alter the path forward. The good old boys club is only a club so long as membership remains exclusive. The silverbacks remain the heavy hitters as long as other members of the group do not rise to their level. Interviewees noted that these dynamics work because women's contributions as well as those from other historically oppressed groups tend to be underestimated or undervalued.

WOMEN AS UNDERVALUED

There are many ways that women's climate change work is undervalued or underestimated. Participants mentioned instances when they or women they knew were not taken seriously or were perceived as less competent than their male peers. They felt that women are often invisible, overlooked, or ignored. This manifests in being treated differently by colleagues, being denied credit for ideas, or being unable to get the same coverage as men even when they are present in science and politics. These obstacles to women's effective or valued participation appear through a women-as-undervalued storyline. Examples of this storyline have already appeared in chapter 5 where interviewees noted instances when their expertise was challenged. Within the women-as-agents discourse, the women-as-undervalued storyline situates challenges to expertise within larger patterns of women feeling that their contributions to climate work are not seen as equivalent to others.

In particular, some women noted that they have been perceived differently from their male, often older, peers. Mary reflected on her experience at the UK Environment Agency by saying, "I quite often found that the communities reacted differently to me as a woman than they did to my male colleagues, in that the farmers don't like being told by the Environment Agency what to do, and particularly if it's a man. But if it was me, they were kind of like 'I don't know what I think.'" Participants attributed these kinds of experiences to their young age, their gender, or both. In these cases, they felt that their expertise and sometimes their position were being called into question. Haley, who works in the United States, felt that some people with whom she interacts in the energy sector fail to take her as seriously as they should. "As a woman . . . especially being a younger woman, I think . . . It's seen kind of like this pet project. Or it's like, 'oh well, that's the thing that she's focused on.' You know, 'that's cute.'" Haley's use of the term "cute" indicates feeling belittled or underestimated in gendered ways. In this passage, feminine things are considered to be cute, including her environmental advocacy.

Because of these associations, some interviewees noted that women feel pressure to change feminine characteristics to become (more) visible in climate change spaces. For example, Brittany argued that this was true in the context of her work for an environmental nonprofit in the UK and also in her interactions with family members. Her father encouraged her to change her debate style in order to be more effective.

> I've learned to not be so passionate, which I think is a bad thing, but also been told not to shout . . . told not to be loud. And had to learn those skills of calming down and to take the emotion away. And there's a pressure to do that to be listened [to] and taken seriously. Which is difficult because it is such an emotional thing to talk about.

In her view, she needed to avoid being seen as emotional in order to be taken seriously in her climate change work. However, she also claims that this is not necessarily a good thing because climate change is an emotional issue. Nevertheless, she has altered her debate style because of her belief that she will not be listened to if she displays passion or emotion.

This example illustrates how gender norms shape assumptions about appropriate behavior from an early age. For many of us, families are a powerful source of information about gender norms. Some participants mentioned censoring or otherwise limiting conversations about their jobs with family members because they either had negative reactions to what they do in the past or were afraid of getting those reactions. Brittany mentioned that her family tended to see her environmental work "as wishy washy and fluffy, and looking after the animals and caring for the environment. Not that it's coming from a place of sound economic sense, or business sense." Likewise, Eva noted that she has family members who view her environmental work in Germany as "kind of like a silly, activist sort of thing to do. It feels like it makes it easier also for them to not necessarily take it very seriously." Brittany and Eva's experiences with family members echo those of Haley in the energy sector mentioned above. They all recounted instances in which they felt that they were not taken seriously. This is likely tied to the feminization of environmentalism—where expressing environmentally friendly attitudes or performing "green" tasks is associated with femininity (Brough et al. 2016; Swim, Gillis, and Hamaty 2020). Many jobs associated with femininity or that are dominated by women tend to be underpaid and less valued than the ones dominated by men. Since environmental action or behavior is often associated with femininity or women's tasks, choosing this as a career can be met with ridicule or at least dismissiveness.

Some interviewees also referred to instances in which women who participated in climate change work or environmental governance were assumed to have lower status or else seen only as tokens or "motivational speakers" rather than as full participants. Haley said she was often referred to as a secretary or assistant in her communications and marketing positions in the solar industry. In her view, people assumed she held a lower-status position due to the fact that she was young and female. Selma specifically discussed women of color and their representation in climate change spaces. She recounted feeling as though people invite women of color to speak merely in order to meet a diversity goal rather than valuing her perspective:

> I think we are still seen as, you know, as cute. And maybe I'm being a bit too critical, but from a feeling, this is definitely something I have felt. And obviously I worked a lot in Africa and also the Vietnamese culture is also very, very patriarchal. We definitely have a lot of prejudice against us, straight away. And they say it, I mean, it's not even hidden. It's like that.

For Selma, patriarchy shapes people's assumptions about the significance of women's work in climate change action. Talia, a nonprofit worker from Germany, reinforced this idea of women of color being used as props or tokens. She argued that women from the global South might be brought into climate change conferences or negotiations as "motivational speakers," but not really valued in the same was as other actors. Instead, they are there to make everyone feel good or progressive. In these instances, participants felt that although women are present, their inclusion was symbolic rather than based on recognition of their expertise.

Working in climate change, like many other issue areas, involves a large number of meetings and conferences. Some interviewees wanted to talk about gendered patterns they notice or experience in these settings and why this might influence how people think about or address climate change. For some participants, these discussions also seemed to be a way of expressing some frustrations or venting about various issues that had been bothering them. Several of them recounted specific stories of women being undermined or belittled at professional events. Kimberly, a nonprofit worker in Germany, gave an account of women being interrupted and talked over at a meeting she recently attended. Both Sharon and Jessie, nonprofit workers in the United States, mentioned the same incident that took place at a board meeting of a large utilities company they had attended for their respective organizations. At the meeting, one of the two female board members pushed to adopt gender-neutral names (i.e., being called "chairperson" rather than "chairman"), and the male board members "treated it like it was kind of cute." Sharon said she went from being encouraged that someone was trying to get people to think critically to being disappointed that it was received so poorly by the people in positions of power.

The way professional business is conducted has long been established in most fields related to climate change. While the format of meetings, the timing of events, and the professional norms concerning who is present have the appearance of neutrality, in reality, however, they influence whether and how one participates, as well as assumptions about who is successful or effective in these spaces. For instance, if an institution has a history of using gendered titles such as "chairman," it reinforces assumptions about who is most likely to be in positions of power. It is noteworthy that two interviewees working for two different organizations both discussed the same event. Each saw it as an example of a woman's attempt to address these gendered patterns of access and position of not being taken seriously by others in the room. Similarly, if an organization holds a large number of events in the evenings and on weekends, many parents may be unable to attend if they do not have accessible childcare. Moreover, if the expectation of an organization is that you will prioritize work at the expense of other things in your life, those who have other commitments will be at a disadvantage. Paige, a nonprofit worker in the UK, reflected on how being a parent would make the kind of climate change work she does difficult:

> Climate change is such an issue of passion and a lot of people who are in it, it is the only thing they care about. . . . And I think that's impossible for people who have multiple things going on in their lives. . . . I do see it all the time in other women who go on maternity, and then have kids, and the dynamic is just totally different. And they inevitably end up in the operational roles because they just can't keep abreast of the strategy, and the policy, and the politics. . . . It's something I've discussed with friends a lot, like, how on earth would you do this with kids?

This passage highlights links between the women-as-caregivers and women-as-agents discourses. The often-intense demands of climate work coupled with the unequal care burden that many women face can result in a barrier to their participation. This sheds light on the fact that climate work cannot be understood without reflecting on how it is situated in larger social, economic, and political structures.

WOMEN'S EXCLUSION FROM LARGER STRUCTURES

Many scholars warn against conceptualizations of agency that fail to recognize how it is embedded in existing structures of power. They argue that agency is "what becomes socially established in any particular historical period as the natural limits of social reality and thence of social practice" (Maclean 1999, 33). Lucy Ford (2003, 124) claims that "any conceptualization of resistance and emancipatory forms of counter-hegemonic agency require [*sic*] a recognition of global hegemony and the dominant agency already inscribed with it." Scholars must understand how power flows through existing structures before we can chart alternative paths through them. A theme that came up repeatedly across the interviews was that women face challenges to participation in climate change activities and that these challenges are rooted in economic systems, political systems, cultural assumptions about "the good life," and other fundamental aspects of daily existence. Thinking about women's agency and obstacles to that agency necessitates reflecting on how gender and other factors reinforce ideas about who should be present in climate change spaces. One of the most important themes to emerge from the interviews is the notion that while there are multiple communities that are deeply committed to addressing climate change, some major features of those communities limit the extent to which this is currently being done in inclusive ways. For instance, women still tend to be underrepresented within the scientific community. Likewise, the activist or nonprofit sphere has a gendered, raced, and classed legacy of mainstream environmentalism (Carter 2018; Curnow and Helferty 2018; Toomey 2018). While there have been important shifts in the academic and nonprofit realms, we must not ignore this history. Simply put, it will be impossible to fully break out of existing patterns of action if we do not identify where problems lie.

Women in Industry and Tech

Numerous interviewees reflected on women's exclusion from the economic decisions that helped usher in the current climate change predicament, including decisions about resource extraction and consumption in the fossil fuel industry. In some instances, this involved drawing a link back to the Industrial Revolution, which profoundly redefined humanity's relationship with the natural world (as discussed in chapter 3). The key figures that are associated with these changes are men, and they are the ones who shaped the revolution's outcome. In the words of Elane, a nonprofit worker in the United States, "I think traditionally . . . women have not been at the table to acquire fossil-fuels, they have not been part of the rampant expansion, they have not been a part of the means by which large-scale electric generation has, in the developed world, come to pass." This is not to suggest that all interviewees think that women's leadership or increased presence would have resulted in a different outcome. As mentioned previously, several participants expressed skepticism that women would have necessarily made different decisions.[10] What they did agree on was that women were simply not the ones making those decisions.

At the same time, interviewees noted that women are also underrepresented in sectors that are supposed to lead the revolution away from fossil fuels, including the "clean energy" sector along with other technological areas. Their responses focused on both the gender imbalances within these fields as well as the dominant perceptions of human-environment connections that accompany these fields. As Ivy, a Canadian academic, observed,

> The sort of dominant discourse of climate change solutions tends to be, in my opinion, quite technocratic. You know, like let's build higher bridges, let's, in agriculture, develop all these new drought resistant seeds and then patent them and protect them . . . and make them very profitable as well. And so, I think that in many ways if we look at the institutions, like in agriculture for example, the large corporations that are putting out these very technocratic solutions to climate extremes, those institutions are very much still male-dominated organizations.

This "technocratic approach," which sees climate change as a problem to be fixed by human ingenuity and intervention, is also deeply connected to a business-as-usual mindset, both literally and figuratively. It fits with capitalist economic models. It also typically absolves people of the need for large-scale behavioral change. Kristy works for an environmental nonprofit in the United States and reflected on the gender imbalance in fields associated with these technological approaches to climate change:

> When it comes to clean energy technologies, it tends to be a very engineering-dominated field, so one could argue that maybe on the mitigation side there are more men technically working on that. It's well known that women are under-represented in the sciences . . . engineering, math, physics—it's more men. So, to the extent that the solutions on the mitigation side would be coming out of those fields, then yes, I guess one could argue that it's probably more men. And particularly, men in developed countries. Which means they're probably predominantly white men.

Kristy was not alone in specifically mentioning clean energy as an area that is lauded for its great potential in addressing climate change while also significantly lacking diversity. According to Hildi, a nonprofit worker in the UK, "The majority of people I know who work in energy cooperatives are men. And the majority of people who benefit from large renewable energy schemes in the manufacturing industries—such as the big offshore wind manufacturing site that is being constructed in Northern England—are men." Likewise, Shea and Sharon noted a similar gender imbalance through their work with clean energy organizations in the United States. According to Shea, "When we look at clean energy enterprises and clean energy sector in general, it's predominantly white men. There's a significant lack of representation [of others]." In Sharon's words,

> Even with the renewable energy industry too, it's men. Underlying it all is the fact that men are most of the executives in any of the big companies that are doing anything revolutionary. I'll see more women in conservation jobs or clean up.

When it comes to climate engineering technologies, there is a vibrant literature on their politics, ethics, and environmental consequences (Biermann and Möller 2019; Jinnah 2018; Jinnah and Nicholson 2019). Some of the most controversial include strategies such as capturing carbon and storing it in the oceans or underground and albedo modification (i.e., intentionally increasing the amount of sunlight that is reflected back to space in order to reduce the amount absorbed by the earth). While these tactics have long been a subject of discussion within scientific circles, they have increasingly moved into more mainstream international climate change debate, as evidenced by their appearance in documents like the Paris Agreement and IPCC reports (Jinnah 2018). What does this increased visibility mean for perspectives on climate expertise? As noted above, some interviewees argued that when people think of solutions to climate change, many bring up technological that which are male dominated. Allison, a US-based academic, said: "I would like solutions to be non-gendered, but for instance, with the predominance in men in STEM fields, it is likely that solutions in terms of technology are going to be mostly created by men. Just because of the predominance of men in those fields." Kit, an academic based in the UK, went further to suggest that not only are more men involved in these fields, but also the emerging technologies are actually designed for them:

> I mean one of the big critiques about energy efficiency and it talks about "resource men" and how these, sort of, smart-energy technologies are, sort of, designed with this ideal "smart man," "resource man" envisaged. Somebody

> who is very technically minded, rational, makes decisions based on economic savings. Not based on the woman who says that I'm not going to put my washing on in the middle of the night because it's going to be creased by the time I come to it in the morning.

Kit argued that women are not wrong when they prioritize something other than economic savings, but they might be seen that way. This reinforces the idea of women being underestimated because standards of evaluation privilege traits typically associated with masculinity.

A few participants noted that in their experience, women often prioritize different aspects of or approaches to climate change than their male counterparts. Deb, an academic working in the UK, said that she saw women as focusing more on "the human elements of security rather than seeing the solution in technological or mitigation through flood gates, barriers, innovation through engineering." She also mentioned that women might tend to be more proactive than reactive. This was echoed by Darcy, a US-based academic, who referred to "technocratic or traditionally male dominated scientific approaches, versus more women-centered . . . approaches that would focus on . . . personal interactions or on-the-ground interventions." Likewise, Janice brought up how these gendered dynamics tend to play out in her own US-based environmental organization and the work it does coming up with solutions to climate change:

> I'm fascinated by the ways in which men and women react to the complexity of addressing climate change and we see this all the time in our office, and these are generalizations, but the men tend to be wanting more technical approaches and to be technocratic about it. . . . Whereas a lot of the times the women will, they want to know that there are solutions, but that they also really want to know how people are going to be impacted and how they're feeling about it.

She also went on to say that in her experience, it is often men who express strong faith in both the effectiveness and salience of technical approaches.

> We will be in meetings where you will hear the women in the organization specifically say we need to appeal to the heart. We need to appeal to a shared sense of humanity. We need to appeal to a sense of community, we need to get mothers involved, they care, you know, about the future of their children. Which

> is not to say that fathers don't, but women will talk more perhaps in that way. So, and then on the male side they'll say, you know, the fossil fuel companies are just going to crush us like bugs if we go all touchy feely on everybody. And what we really need to do is prove the promise of the clean energy economy. And here are all these economic studies that demonstrate that the clean energy economy is inevitable, and we're in a clean energy revolution. And here's all this data and all these statistics that tell us that were well on the path. And we just need to use information and facts and figures.

By contrast, Nora, who works in the nonprofit sector in the UK, argued that both men and women are "sucked into techno fix approaches" to climate change. These kinds of paths are attractive to many because they offer the promise of addressing climate change without the need for significant behavioral change. Feminist scholarship has outlined multiple ways that geoengineering intersects with gender, including the demographics of those pushing the method, the overall discourse of control involved in geoengineering, the design of the particular technologies, and the populations who will benefit most from geoengineering (Buck, Gammon, and Preston 2014).

Annise, another US-based nonprofit worker, wondered why more men in positions of power do not try to benefit from the transition away from fossil fuels.

> A lot of times I'm just thinking at home like, "Alright white men, just get on the renewable resources, you know, like, there's going to be a lot of money in those things. Just get off oil. Get off all of your dirty fossil fuels and make a bunch of money in the next, you know, exploit the next industry, whatever. But move on."

Assuming that it will be white men who will be in a position to take advantage of the next revolution in industry, Annise added that their motivations could actually be useful. "I think we need some greed and some power hunger [to], kind of, generate the industry we need that will create a cleaner economy and environment." In this view, a goal of immediate sustainability is prioritized over long-term shifts in the structures that contribute to climate change. This trade-off was mentioned by multiple participants working in the environmental nonprofit sector.

Women in Environmental Organizations

In the aftermath of the "Environmental Revolution" discussed in chapter 3, environmental organizations with global reach were the ones most successful in attracting funding and cultivating strong networks. A good deal of this depended on board members being part of established power systems that could amplify their position (Ignatow 2005). Thus, professional environmental organizations came to resemble other male-dominated professional entities as they grew more formalized with paid positions and formal bureaucracies (Stover and Cable 2017). Historically, women have been both absent from leadership roles and underrepresented among support professionals (i.e., lawyers, scientists, and the like) in most environmental organizations. At the same time, women have comprised a large percentage of their members (Seager 1996; Taylor 2016).

While participants acknowledged the widespread participation of women in environmental and climate change organizations in general, they also noted a lack of women in top positions. Swati, who works for an environmental nonprofit in the United States, wondered how the predominance of older white men in well-funded organizations "impact[s] the analysis of the issue, and the solutions that are presented, and leadership on the issue" of climate change. Those in leadership positions may not see environmental issues the same way as communities with a different set of experiences.

Some interviewees also described events such as conferences or industry presentations where they still saw a large number of "men in suits." This picture was often held up as the contrast to the women-dominated or at least gender-balanced organizations they worked for. These "men in suits" (sometimes "fancy men in suits," sometimes "white men in suits") were discernable at solar industry conferences, environmental advocacy awards events, UN panels, and even photoshoots. Paige, a co-director of a climate change organization in the UK, recounted that during a campaign her team delivered with another, well-known global environmental organization, she was she was "basically shoved out of photos." The male director of the big environmental organization told her that "the thing is, photographers they want photos of men in suits." In her view, this is an outdated picture of environmental action. She asserted that "we have seen plenty of photos of men in suits standing in front of things." However, the fact that the male director held on to the assumption that men in suits best represent environmentalism likely speaks to larger societal expectations about the face of "professional" environmentalism.

While some might expect environmental nonprofits to be progressive on matters of gender due to the large presence of women within them, many participants argued that this is not the case. According to Janice, who works for a climate change organization in the United States:

> So, all of the big gender sexism challenges that we have in society . . . exist for sure in the climate space. Without a doubt. And I, so I was at Yale in the seventies, which Yale was just changing from being all male to being co-ed, and the journalism space was also extremely sexist. And the high-tech space which I was in for seven years was also extremely sexist. So, I wouldn't say that the climate space is any more or less sexist and gender biased, but it's the same old shit.

In her view, climate change organizations exhibit some of the same gender dynamics at play in other areas she has worked in. Patriarchy remains an ever-present obstacle to truly equal participation at present despite the large presence of women in climate change or environmental organizations. For her part, while Swati believed that the composition of organizations is definitely changing to become more diverse, she thought that "it's changing a lot slower than most people would like." She explained that her environmental organization currently is and historically has been led by white men. About a year prior, they decided to engage outside consultants for a diversity assessment of the organization because they understood that they "needed to be much more reflective of the community" they are serving. There was also reflection on how broad movements can facilitate change. In her words, "if you looked at our staff and board it's not reflective of that at all, but then also, looking internally at our culture too, that, you know, we still act in very much in a white male heterosexual kind of dominate frame, so how do we make ourselves be more inclusive in different ways?" She noted that the diversity assessment made members of the organization confront uncomfortable trends in how they interacted with each other. They collectively reflected on issues such as how the organization engages with the community, but also more specific interactions like who cleans the office kitchen and brings in the birthday cards for colleagues.

This speaks to a trend of women and men performing different tasks in environmental organizations. In Sharon's words,

> Every nonprofit is going to have communications, development, organizing . . . and then you will get people coming into the technical analysis fields. For the most part the data analysis people are men. The technical conversations you find yourself in [are typically majority men].

She argued that most of the "feminine" tasks of an organization (i.e., communications, education, outreach) tend to be performed by women. This is the case for her US-based clean energy organization and many others she is familiar with. Similarly, Aubrey, who works for a climate change organization in Germany, indicated that women's presence tends to be seen more in areas such as climate change education rather than lobbying or tasks directly related to climate science. She mentioned one of her friends in the lobbying industry feels the pressure to "prove herself" in this profession:

> I know that she faces a lot of issues of . . . you know, when she walks into the room, having first to make lots of very good scientific statements before people will accept what she's saying because she's a young woman. . . . She really feels this barrier of, you know, being a woman. Having to, kind of, prove herself first before people trust what she's saying. Whereas she doesn't feel that some of her maybe older male colleagues have to fight that barrier.

In addition to highlighting the "barrier" of being a woman, these observations are consistent with the women-as-undervalued storyline.

A few participants also pointed out how the large presence of women in environmental advocacy intersects with other issues of diversity and inclusion. Heather, a US academic, pointed out that women tend to be found more at lower levels than at the top. She argued that there is likely more space for women's voices, including indigenous women, to be heard at lower levels because of existing power dynamics. June, who works at a US environmental organization, observed that achieving gender balance will get an organization only so far if it truly has goals of inclusion and transformation:

> I feel like also in the environmental movement we've been talking a lot about how we've been so historically white, but and often we don't really talk about the

> fact that, we don't address gender that often because, especially in our organization, we are led by all women, and our two presidents, our CEO, one of our board chairs, they're all women, and our staff is like 70 percent women. So, it's not something that we talk about a lot. But . . . even though we are led by women, there is still this patriarchal and hierarchal nature of our organization that I think is probably reflected in a lot of environmental nonprofits that are currently being led by women but have historically been set up by men.

June argued that for organizations to embrace collaborative processes and avoid exclusionary top-down decision-making, they need to think about the composition of their staff as well as how the organization is structured and interacts with the community. In the same vein, Annise commented that "we can't just win as a bunch of predominantly white middle-class, middle- to upper-class outdoorsy people." She reflected on the need for environmental organizations such as hers to build diverse coalitions with people working toward similar goals. Feminist scholars have illustrated that it is not only the composition of institutions that is important for equitable outcomes; rather, the process of decision-making and the behavior of actors also matter a great deal (Magnusdottir and Kronsell 2015).

Though interviewees found it encouraging that women make up such a large percentage of the environmental movement, they also voiced frustration that they are still underrepresented in "masculine" areas of organizations and in top positions. Those areas seen as "masculine" often overlap with the STEM fields discussed in chapter 5, including those involving technical and data analysis. Participants also acknowledged that white men dominate the leadership of most major, wide-reaching organizations. They largely failed to mention the many gender and climate change organizations or networks that have been active for years in multiple climate spaces, including GenderCC, WEDO, and many other national and global environment and gender organizations.[11] Their reflections about the environmental movement focused mostly on their own organizations in the case of nonprofit workers or well-known organizations in the case of those outside this sphere.

IMPLICATIONS OF THE WOMEN-AS-AGENTS DISCOURSE

This discussion of environmental organizations illustrates one of the major features of climate change within the women-as-agents discourse—climate change is a space of work. This work involves politics, industry, tech, nonprofits, and more and requires navigating existing norms and power structures. Climate change is also impacted by human actions such as consumption or organizing—sometimes positively and sometimes negatively. It also forces people to consider the future. Will our society transition to clean energy? A green economy? Transformational perspectives and norms on human-nature connections? Climate change forces us to consider a future where the status quo is impossible, but we disagree on what or how much change is possible.

Where are the women in the women-as-agents discourse? Interviewees saw women participating in climate work or action at multiple levels. They were in environmental organizations, at city council meetings, and in classrooms. These women are consumers and climate bad asses, motivational speakers and leaders in the environmental movement. The women-as-agents discourse, like the women-as-knowledgeable discourse explored in the previous chapter, pushes back against pervasive depictions of women as passive victims of climate change. It highlights choices women make to get involved in environmental and climate change action, as well as choices women make that might contribute to environmental change. Participants reflected on their own agency, as well as that of women in general. In the case of their own agency, they outlined specific actions they take and the reasons behind them. They explained their passion for getting involved as well as the toll that the work can take. In particular they often identified women's strong presence in environmental work as a source of pride and inspiration. In general, there was a greater tendency to note women's agency in tasks associated with advocacy or adaptation rather than mitigation.

The discourse also features a great deal of reflection about where women are absent. Participants pointed to the underrepresentation or absence of women from STEM fields and the fossil fuel industry, as well as clean energy companies, and politics. They also noted their tendency to be missing from top positions in environmental organizations. They identified some barriers to agency, including the composition of the worlds of activism and the academy. In order to think of women as agents, it is essential to reflect on what keeps people from being agents or else in some way negatively influences their agency.

Participants offered multifaceted portrayals of women's agency. They reflected on collective action that women participate in as well as their everyday activities. Reflecting on agency and "everyday activities" reminds us that people make choices that shape environmentalism, but that these choices are simultaneously shaped by various forms of power in the international system. According to Sherilyn MacGregor (2021, 56), a great deal can be learned from approaches that regard "everyday life not as a euphemism for 'what ordinary people do,' but as a site of 'production, reproduction, exploitation and domination.'" Interviewees acknowledged that thinking of women as agents means identifying instances of powerful climate action and also times when women's choices contribute to the current environmental predicament. Participants' use of the women-as-consumers storyline illustrates this complexity. They used the storyline to reflect on their own consumption patterns, assess gendered patterns of household labor, and critique essentializing assumptions about women's overconsumption. At the same time, participants who used this storyline overwhelmingly associated women with household consumption, including green consumption. They are not alone in this association as companies themselves tend to target women with eco-friendly products and marketing (Hunt 2020). Assuming or encouraging women's green consumption can have multiple implications for society, including adding to women's already unequal domestic responsibilities, as well as potentially encouraging men to view environmentally friendly behavior as feminine. Some existing research identifies an implicit cognitive association between the concepts of greenness and femininity. The work finds that the association can affect social judgments and self-perceptions and lead men to be less willing to engage in green tasks linked to women's work (Brough et al. 2016; Swim, Gillis, and Hamaty 2020).

Whether consumption should be understood at an individual level is another issue. While the notion of individual agency, or "power to," captures the imagination,[12] individual efforts at sustainable consumption are unlikely to be sufficient to effectively tackle climate change (DeSombre 2018). They are also cost-prohibitive for many people and can result in classed perceptions of who is environmentally virtuous and who is not (Scerri and Magee 2012). Participants who used the women-as-consumers storyline overwhelming described middle-class women like themselves without reflecting on consumption as embedded in other systems of power, including class or race.

The majority of participants who brought up consumption saw it as just one aspect of climate change. Lily's comment about "silly internal struggles in the supermarket" indicates that she doesn't regard sustainable consumption as the final solution or the most important aspect of her climate change work. Likewise, Selma's remarks about consumption framed her individual challenge in larger structures. She wondered how she, as a working, single mother, is supposed to make the greenest choices with limited time. This is an instance when the women-as-agents discourse and women-as-consumers storyline make a useful contribution to thinking through climate change action—by conceptualizing women's agency as located in larger economic, political, and social structures—but more reflection is needed about this embeddedness.

Numerous participants who highlighted women's place in domestic consumption specifically linked this role to the women-as-caregivers discourse outlined in chapter 4. They discussed women's responsibilities caring for families and households as a reason why they tend to have a larger say in purchases of food, appliances, and the like. In one participant's view, "Women are the ones who are playing that role of being the caretaker for the family." The idea of "playing a role" indicates that there are other possibilities, but this is what society expects of them—this is their position as the script of gender norms has been written. None of the participants used the feminized "shopaholic" trope that one participant found so frustrating, but instead tended to point out why they personally consume the way they do, or why women's consumption might lead to greater awareness of climate change impacts or even opportunities for greener consumption. Environmental scholars and activists have loudly called for reductions or changes to consumption but fail to adequately interrogate the gendered aspects of why people consume the ways we do (Detraz 2017b; MacGregor 2021). The women-as-consumers storyline features reflections on how consumption is tied to dominant images of both masculinity and femininity. In this way, participants voiced a facet of consumption debates that scholars in fields such as global environmental politics have been slow to address.

That being said, there tended to be a good deal of homogenizing about "women" across the women-as-agents discourse. The sites of agency as well as patterns of restriction that participants identified often followed dominant portrayals of women's environmental action. Seema Arora-Jonsson (2011) describes this as the assumption of women's virtuousness in the global North, contrasted with women's vulnerability in the global South. The assumption is that white, northern women are "more sensitive to risk, more prepared for behavioural change and more likely to support drastic policies and measures on climate change" (Arora-Jonsson 2011, 746). In her view, these kinds of assumptions are reinforced by reports on gendered differences in ecological footprints that have been cited extensively by those working on climate change and gender (Johnsson-Latham 2006). It is essential that women's agency is recognized and supported in order for climate change efforts to be both effective and equitable. At the same time, it is important to unpack discussions of women's agency to understand how they might unintentionally reinforce the very structures that have made women's participation in multiple facets of climate change work difficult. For scholars such as Arora-Jonsson, painting women as either virtuous environmental saviors or hapless environmental victims can have its own set of negative implications. These tropes can, and are, deployed with little interrogation about power distribution across existing social, political, and economic structures. Women become a category of people "to bring in" either because of their virtuousness or victimhood. This "bringing in" potentially avoids asking questions of why women's participation in various areas of environmental protection has been lacking in the first place, including questions of how and why their agency has been limited. This is in no way to suggest that women's inclusion should be shunned or limited, but rather to ensure that this inclusion allows for their full selves to be present and not the imagined saviors or victims that may be expected of them.

Additionally, how do assumptions about women's agency and action influence the tasks they end up performing? Does assuming that women care more for families and communities mean that they will continue to play a larger role at that level? Is women's participation in a local environmental organization somehow more natural than their involvement in international climate change negotiations? Some participants touched on these ideas in our conversations. Brenda, a nonprofit worker in the United States, mentioned that women are "supposed to be, kind of, doing the good work." Women are expected to be active and be caretakers, while men are presumed to be the breadwinners. These gendered stereotypes impact who feels the emotional weight of environmental protection along with how some environmental work is compensated. A few participants noted that there might be more women active in environmental work, particularly at lower and middle levels, because it is poorly paid and undervalued. This is connected to the notion that "doing the good work" is its own reward, or that fair compensation is more important for men since society expects them to be providers. If that is the case, using essentializing frames to discuss women's climate change actions reinforces these assumptions. Even if participants overwhelmingly indicated that they get emotional satisfaction or other nonmonetary rewards from their climate change work, it is still problematic if the work is undervalued—either economically or in status, something that frequently occurs with women's labor across the globe (Benería 1995, 2003; Coffey et al. 2020; Elias and Roberts 2016; Waylen 1997). When the women-as-absent or women-as-undervalued storylines were used, it was often in reference to social norms. Women were not formally blocked from taking part in activism or scholarly endeavors. Rather, participants felt that they or women like them were overlooked. There is great deal of scholarly evidence that women's contributions to environmental action are frequently misunderstood or marginalized (Kennedy and Kmec 2018). This is particularly the case as women's household or community tasks are depicted as small-scale and less important than some other climate work (Whetung 2019). Additionally, simplistic assumptions that women might be particularly well-suited to environmental tasks because of an essentialized vision of their "naturalness" persist (Gonda 2017).

Large numbers of women play significant roles in everyday activities and collective action related to climate change. A large contingent of youth climate activists has young women leading the charge (Eve 2020). The most visible of these in the global media has been Swedish activist Greta Thunberg. In August 2018, she started a school strike for the climate outside the Swedish Parliament building. That movement has spread to cities around the world and inspired school strikes by children in numerous countries along with the large-scale global climate strikes in the fall of 2019. Thunberg took her message of climate action to numerous seats of political and economic power—including the United Nations, the European Parliament, the World Economic Forum, and governments in multiple capital cities. She was named *TIME* magazine's Person of the Year for 2019 and has been the subject of an outpouring of news stories, including being pictured in *TIME* with the headline "Next Generation Leaders." Although Thunberg has received the lion's share of attention, she is just one among a number of youth climate activists in public eye. In fact, there are multiple young women of color leading the US youth climate movement, including Jamie Margolin, Nadia Nazar, and Madelaine Tew, founders and executive directors of the Zero Hour movement, and Isra Hirsi, executive director of the US Climate Strike (Burton 2019). It is essential to avoid limiting these and all other women's agency with relatively narrow frames of the kinds of roles that they might be particularly well-suited to play. Reflecting on women as agents requires assessing where and how women have made important contributions to climate action along with the consequences and challenges to this action.

Women's climate action is sometimes met with distain and derision, especially when their message challenges the status quo. There have been numerous commentaries detailing the intense and charged language used to attack female climate figures (Gelin 2019; Moore 2019; Raney and Gregory 2019), including Thunberg, Canada's former minister of environment and climate change Catherine McKenna, and US Representative Alexandria Ocasio-Cortez. This trend is consistent with academic work that points to substantially significant linkages between climate denial and gender norms (Anshelm and Hultman 2014; Jylhä et al. 2016; Krange, Kaltenborn, and Hultman 2019). For example, research on climate skepticism in Sweden by Jonas Anshelm and Martin Hultman (2014, 85) notes that the phenomenon "can be understood as being intertwined with a masculinity of industrial modernity that is on decline." In their view, climate sceptics adopt this position in part because it saves "an industrial society of which they were a part by defending its values against ecomodern hegemony." Rejecting climate change and belittling climate activists, especially young female activists, serves as a coping mechanism in a period of transition when many members of privileged communities feel that their privilege is eroding. The increased awareness of Earth's climate reality means the actual and potential shake-up of power structures, which could mean that some who have benefited the most from current social, economic, and political structures will lose this privileged position. This might help explain why multiple public figures in the global North have made sexist and ableist comments about a teenage climate activist such as Thunberg (Gelin 2019; Moore 2019). She may be seen as representing a future in which young girls are not blocked from participation by their gender, age, or social position.

The women-as-agents discourse highlights the positives of recognizing and expanding women's environmental action, but it also necessitates acknowledging the negatives that sometimes accompany this participation. The previous chapter mentioned that some climate scientists have experienced gendered or sexualized harassment when their research gets picked up by mainstream media or climate skeptic outlets (Johnson et al. 2018; Ogburn 2014; Waldman and Heikkinen 2018). Likewise, women who act as "environmental defenders" have been subjected to threats of violence, threats against their families, and murder (Glazebrook and Opoku 2018). While environmental activists have historically been at risk of being criminalized or experiencing violence or human rights abuses (Global Witness 2019), these risks take on a particular cast for women as they are often specifically gendered. Dalena Tran and colleagues (2020, 1190) explain that "women activists are often delegitimized based on their gender. Trying to enter public, political spaces for debate turns them into targets for multiple forms of violence. Murder is the most visible of these in environmental conflicts, but all threats to women defenders are difficult to document owing to intersecting marginalities and stakeholder interest in covering up abuses, as well as the fact that literature on violence in environmental conflicts does not separate data by gender." This means that thinking through women's agency must necessarily entail reflecting on not only obstacles to their participation, but the consequences they may face for this action.

Who are the women most frequently included in the women-as-agents discourse? Participants disproportionately focused on women in the global North. Participants mentioned women from the global South as agents, but they often saw them as resource users. They did not talk much about women from the global South buying cars or attending protests. Women from the global South were mentioned as advocates for climate change action, but typically these were indigenous women, not employees of professional environmental organizations.[13] Likewise, the categories of marginalization discussed in chapter 3 (i.e., race, age, disability, class, and place) did not feature heavily in reflections on agency even when interviewees described environmental action in the global North. The exception to this is in reflections on the overrepresentation of white men within existing power structures.

That, in turn, led to observations about women's absence from or undervaluation in multiple forms of climate change action. These storylines call attention to marginalization mainly in professional spaces. It highlights that simple calls to "bring in more women" might ignore the significant number of women who already exist in climate change work or result in bringing more white women into climate spaces while doing nothing to address racial or other imbalances. For instance, we might ask why it is that Thunberg, a white Swedish activist, has garnered so much more global attention than youth activists of color. The fact that Thunberg has risen to such prominence makes her an unfortunate target of scorn and derision, but it also raises questions about why the global media has tended to pay more attention to white activism and disregard movements led by people of color. It is essential to listen to those activists and academics who have explored why environmentalism in the global North tends to be very white (Carter 2018; Curnow and Helferty 2018).[14] Suzanne Dhaliwal (2015), the co-founder of the UK Tar Sands Network, argues that "if you were to trust what you see in the UK media you would think that climate change is a white issue that speaks to and is populated by one demographic alone." This speaks to the portrayal of global issues such as climate change, but it also relates to the composition of mainstream environmental organizations. A 2014 study that examined diversity within US environmentally focused organizations found that after close to five decades of diversity goals, gains in representation were uneven (Taylor 2014).[15] The greatest gains were made in the area of gender, but white women far outnumber women of color in these organizations. There is a significant lag in achieving racial and ethnic diversity overall in these spaces, as the percentage of people of color on the boards or general staff of environmental organizations was less than 16 percent, and less than 12 percent of the leadership positions. Additionally, the members and volunteers of environmental organizations are predominantly white. Furthermore, a study of over 2,000 environmental nonprofit organizations in the United States found that the percentage of nonwhite staff and board members remains low, while fewer organizations are now voluntarily reporting their diversity statistics. The study also found that men make up less than half of the staff of the organizations but occupy 62 percent of board positions (Taylor 2018).

This finding is in line with the fact that interviewees' professional environmental organizations largely tended to have a history of white men in positions of power. While this trend was shifting, the overrepresentation of white voices was still raised as a problematic element of climate work. The same structures that have resulted in gendered patterns of participation and representation in climate action are also raced and classed. Thus, while women tend to have fewer seats at the leadership table than men, white women are still more likely to occupy those seats than women of color. The whiteness of the environmental movement, along with whiteness in science, came up multiple times across the interviews. Participants commented that it is largely white people in positions to tell us what the world is and how it should look in the future.[16] Reflecting on women as agents affords us the opportunity to assess where and how women have made important contributions to climate action along with the consequences and challenges to this action.

7 CONCLUSIONS: WHERE DO WE GO FROM HERE?

I am writing this conclusion in the midst of a global pandemic. The COVID-19 virus has laid bare many of our incorrect assumptions about human superiority over "nature." It has illustrated that border walls and checkpoints do not match our globalized world. It is a world in which public health and environmental crises cannot be solved by uncoordinated action by inward-looking states. The pandemic has resulted in unprecedented economic upheaval and political uncertainty. For decades, environmental scholars have warned that true sustainability will require not only changes in human behavior, but fundamental shifts in our ways of thinking and speaking about environmental problems. Over the course of a few months, as country after country issued shelter-in-place or lockdown orders and worked to designate who and what was "essential" or not, debate about our future began. I can think of no better time to reflect on something as important as climate change and how it is understood and framed. Hopefully, what society takes from this COVID-19 crisis is a willingness to critically engage with our world and think through how we can effectively and justly address climate change.

The previous chapters underscore that a change in perception is needed in order for people to fully understand the importance of connections between gender and climate change. To this end, the preceding chapters have used a feminist constructivist approach to outline four complex, overlapping discourses specifically focused on women's characteristics and roles in climate change. From depictions of how climate change will act on women's bodies to a celebration of their essential expertise, the discourses show the considerable variance in women's position in climate change. Participants reflected on their own experiences working in specific climate change spaces, such as industry conferences, meetings at their nonprofits, classrooms, and the halls of Congress. They also discussed the experiences of "women" as a general group that has been historically marginalized but has a great deal of capacity and ability. Taken together, these four discourses represent women as interested in and capable of contributing a great deal to our climate struggles. They portray women as active but often blocked. They point to them as vulnerable because of gendered marginalization but not weak or passive. The discourses describe a multitude of roles that women currently play and should play in the future. They also offer specific insight into interviewees' perspectives on both gender and climate change individually and together. Providing the perspectives of women who are currently working "in climate change," the interviews serve as a way to explore the dominant discourses that run through climate change spaces, something that is necessary to understand if scholars and practitioners hope to shift our discourses to be more inclusive and avoid essentialization.

WHAT DOES THIS BOOK SHOW US ABOUT CLIMATE CHANGE?

Each of the discourses explored in this book features a particular understanding of climate change. These range from a phenomenon that exacerbates vulnerability and inequality to a scholarly specialty that showcases the passion and agency of women. Mary E. Pettenger (2007, 5) explains that this is not surprising.

> Ask ten people how to define climate change, its causes and effects, and you will get ten different answers. The language used to discuss and describe climate change is often value-laden as the terms employed have different meanings depending on who is discussing the topic and why. What is clear is that the meaning of climate change is defined in social settings.

This socially constructed nature of climate change means that how we understand and refer to it is connected to our identity and context. For example, participants spoke about debates and initiatives on "how to tackle climate change" as a global phenomenon. They asked whether we can change peoples' minds "around the issue of climate change"—signaling the contested nature of climate change as a political and social issue. They reflected on people's ability "to adapt to climate change or manage in a world with climate change." And they expressed pride in or frustration about "climate change work." Thinking about these various ways of talking about climate change is helpful for understanding why fierce debates about the concept continue across the international system when there is such a high degree of consensus among the scientific community regarding some of the most basic facts about it (Cook et al. 2016).

The book's introduction featured some reflections on climate change and how it has become such a prevalent topic of scholarship and policy debate. I referred to a description of climate change as a "super wicked problem" to highlight some of the intense political challenges facing climate governance (Levin et al. 2012; Rittel and Webber 1973). Additionally, climate change is an immensely complex issue. Understanding and addressing it involves a variety of sectors, as well as many people doing different kinds of work. In the words of one participant, German nonprofit worker Eva, "What's challenging but perhaps also enriching is that climate change . . . impacts too many different fields. You need to work with a lot of different people to actually make a change." Yet those people may view paths forward differently due to their disciplinary backgrounds or because they have different climate change experiences.

Participants outlined several ways this climate change complexity can make action challenging. For instance, UK nonprofit worker Yvette mentioned her frustration that her work was such a small part of what the government is doing. While it dedicated a relatively small pool of money to climate change programs, the vast majority of funds went to sustaining and reinforcing our existing ways of doing things. She argued that there remains an unwillingness to confront the overall shifts in political, economic, and social structures necessary to effectively address the problem. All available evidence suggests that structural transformation is needed in order to avoid a climate crisis. Scholars have put forward various strategies for devising alternative frameworks and governance approaches to get us past our climate gridlock (Mitchell and Carpenter 2019; Underdal 2017). Multiple interviewees similarly suggested that climate change requires fundamental shifts in society that are not yet occurring. Several argued that rethinking gender and climate change can perhaps encourage us to think more explicitly about the social components of climate change and that this will aid in grasping the scale and direction of necessary change.

Some participants also posited that a challenging aspect of their work is to get people in the global North to recognize climate change as a current, dire situation. This was expressed particularly often by interviewees from the nonprofit sector. An example comes from Marie, who works for a German environmental organization. For her, "One of the biggest challenges is that it remains for most of the people I've worked with, a somehow distant thing, phenomenon, and therefore it makes it hard to, to make it visible." She further noted that this intangibility of climate change for many people makes effective climate messaging difficult. At the same time, what does it say about us that something has to be immediately, personally threatening or risky before we care about it? Could we get people to care about climate change even if they are not negatively impacted by it today?

Feminist environmental scholars, among others, have long argued that humanity needs to rethink our relationship with the environment, which necessitates rethinking our relationships with each other as well (Butler 2020; Sturgeon 2008). Rather than being solely concerned about climate change's negative impact on each person, scholars and practitioners should ground our thinking in recognizing the intricate connections between people and the environment, as well as among different communities on the planet. Perhaps thinking in terms of groups of people might help in this endeavor. I argue that the women and climate change discourses outlined in this book enable us to think more critically about various lived experiences of climate change. While focusing on women as a category can potentially lead to essentialization if not approached carefully, it also encourages us to think of the many ways that human beings are similar. We can, for instance, use the discourses as a starting point to reflect on ways that patriarchy and capitalism influence the lives of people and the planet. They also allow us to realize how the positions and lives of women are similar and how they differ. We can think through ways the climate change experiences of middle-class women in the global North, including me and most of my interviewees, diverge significantly from that of other communities (Arora-Jonsson 2011). This kind of critical engagement with human experiences of climate change requires the input of multiple voices from numerous backgrounds, geographic spaces, and lived and professional experiences (Green and Hale 2017; Javeline 2014; Keohane 2015). Accounting for different voices subsequently renders cross-disciplinary conversations about climate change as well as gender possible.

These kinds of conversation necessitate understanding the opportunities as well as obstacles within existing climate change frames. Sherilyn MacGregor (2010) argues that discourses such as ecological modernization, environmental security, green duty, and population control have had a significant influence in climate discussions within the global North. However, rather than provide a transformational, sustainable, and just path forward, these have often served to reinforce existing economic, political, and social structures. Likewise, Joanna Wilson and Eric Chu (2020, 1086), argue that a discourse of the "green economy," which cast climate change "as a problem of science urgently requiring technological advancement and market fixes, means that solutions lie firmly in the domain of men and masculinist priorities." Thinking of climate change as only a scientific problem leaves questions of power and discourse unexamined.

At the same time, how can we recognize the importance of scientific data along with climate change mitigation and adaptation technology while also reflecting on how exclusively paying attention to these may be detrimental to the goals of equity and sustainability? It is essential that we are open to critiques of the academic and advocacy spheres, while also recognizing their importance. We live in an era where climate skeptics and those with a motivated interest in fostering climate denial would love to use a critique of science to discredit all the good work done by climate scientists. For instance, some climate skeptic blogs have published about gender discrimination in the Intergovernmental Panel on Climate Change (IPCC) seemingly in an attempt to shame the institution. It is not my wish to add fuel to that fire. However, it is also important to ask challenging questions so that we can more directly assess the consequences of using a particular climate change discourse. Likewise, we need to reflect on the current state of participation by marginalized groups in climate spaces. In the case of my interviewees, many of them expressed pride in their fields (i.e., atmospheric science, climate change organization, and the like), but also noted how the fields could do a better job of fostering an environment that ensures multiple groups of people feel valued.

Along these lines, we must interrogate how our socially constructed views of science intersect with existing power dynamics across the international system. For instance, we might ask whether current approaches to expanding our idea of "science" always help alter views on women and scientific expertise. As noted by one of my interviewees, Kit, this may not be the case. She used her experience researching community-based environmental projects in Nepal to argue that attempts to draw more people into "science" and environmental sustainability approaches can sometimes unintentionally perpetuate existing power distributions within society.

> [They] sort of served to reinforce [the women's] ideas that the local elite, who are basically the men, they were the knowing ones. They were the powerful ones. "What do we know? We just serve the tea." Because I was sort of looking at the way that it was all very embodied—very sort of situated experience that the women actually took part. And some old women particularly, they took part, but they really just brought the tea along and tied some ribbons to a tree. So, their experience of the projects was very different from the men who actually had the tape measures and were recording numbers. So, it's that kind of embodied experiential nature of what they were doing that led them to have very different experiences. And I guess that maybe, that's true of so many different community-based projects specifically around climate change.

Kit identified gendered tasks that reinforce assumptions about who is knowledgeable and who is a valid participant (i.e., those who did the measuring and recording) and who is not (i.e., those who served the tea). Women's role as teamaker was not regarded as work that made the same kind of contribution to the project, even by the women themselves. Programs that are based on harnessing local knowledge or engaging communities in environmental management are typically motivated by the goal of widening our understanding of expertise and getting "buy in" from local people on conservation schemes.

However, what is not always incorporated into these projects is the fact that local knowledge, like all knowledge, is gendered. It is shaped by gendered expectations of the roles and responsibilities that people are supposed to fulfill. Thus, it is shaped by the same social, political, and economic structures that assign less value to characteristics associated with femininity (Enloe 1990). These reflections are supported by other academic work that points out the unintended consequences of some community-based environmental programs that fail to take power dynamics like gender, class, caste, and race into account (Agarwal 2009; Staddon, Nightingale, and Shrestha 2014). If we fail to reflect on these when strategizing about climate change action, women and other marginalized groups will continue to be constrained.

Likewise, the move from fossil fuels to clean energy technology appears to be following similar patterns of reinforcing ideas of who is knowledgeable. Added to this are the ranks of geoengineers working on new strategies of climate change mitigation. Relying on techno-fixes raises workers in some fields to the position of saviors. According to Sherilyn MacGregor (2017a, 19), "At the uppermost levels of global climate politics, white-Western-male geoengineers are celebrated as modern-day Baconian Supermen who can harness the powers of techno-science to control the very weather." There is evidence to suggest that women are often skeptical that technical solutions such as carbon capture and storage or further development of biofuels are sufficient or preferable means to address climate change (Hemmati and Röhr 2009). These gendered patterns require more research, but they likely relate to larger trends of technology being associated with a "masculine" approach to problem-solving. Meg, a nonprofit worker in the United States, raised the idea of the hero trope and masculinity in her response to whether climate change is a masculine issue area:

> I guess I can see in sort of our gender constructs of masculinity how maybe the thought is that men are more inclined to want to, I don't know, save the world, or like, you know, take on this big feat of climate change and do it themselves. . . . I think it's just historically the gender construct that we've created around men. . . . The first superheroes we really see—Batman, Spiderman—are all men. And we've always sort of perpetuated this idea in our society that men are the ones who take care of their women and, like, pay for their dinner, and carry them over puddles, and let them borrow their sweater.

Meg's reflections on our socially constructed notions of heroic behavior reiterate MacGregor's idea of men being cast (or casting themselves) as the saviors of our climate change tale.

Feminist environmental scholars have routinely been among those raising concerns about the reliance on technological fixes to environmental issues. One critique is that society is easily dazzled by the promise of these approaches without engaging in a thorough examination of the social processes that make them attractive. According to scholars like Val Plumwood (2002, 8), "Technofix solutions make no attempt to rethink human culture, dominant lifestyles and demands on nature, indeed they tend to assume that these are unchangeable. They aim rather to meet these demands more efficiently through smarter technology." One interviewee, Elane, referred to work that her US-based environmental organization has done on the potential perils of geoengineering. She claimed that these are "false solutions" to climate change because they don't necessitate making large-scale changes economic and political systems in order to address greenhouse gas emissions. She likened them to filling up a bathtub with a hole in it. This perspective finds some empirical support in the scholarly literature. For instance, Marijn H. C. Meijers and Bastiaan T. Rutjens (2014) find that when people are presented with frames that portray science as rapidly progressing, it has a negative effect on environmentally friendly behavior. They attribute this to the way that science reinforces an idea of the world as orderly, with threats to that order being addressed by an external actor. This means people feel less motivation to play a personal role in environmental protection since we expect "science" to handle it.

We must also consider the role that techno-solutions to climate change might have in reinforcing existing economic and social structures, or possibly even making things worse in terms of gender equity or social justice. The Center for International Environmental Law (2019) has argued that geoengineering approaches to climate change mitigation can reinforce the dominant position of the fossil fuel industry. Its 2019 report underscores the role of fossil fuel producers in developing, patenting, and promoting important geoengineering technologies in ways that result in prolonged reliance on fossil fuels. Despite these potentially negative aspects of techno-solutions, they continue to be praised by international leaders. In 2019, German chancellor Angela Merkel mentioned "the role of technology and innovation, particularly in the field of energy, but also in the field of energy savings" as important elements in reaching climate goals. Likewise, British prime minister Boris Johnson described technological advances "making renewable energy ever cheaper, aiding our common struggle against climate change" (quoted in Bershidsky 2019). As long as techno-solutions are touted on the world stage as a great hope for avoiding a climate disaster, we should reflect on how these approaches might represent business as usual far more than we might think.

Likewise, cross-national studies illustrate that approaches to climate change that are rooted in existing development paradigms can exacerbate gender inequity rather than mitigate it.[1] A 2015 study from Nepal, for instance, finds that agricultural adaptation strategies can adversely affect women in a number of ways. Practices such as adopting high-yield cash crops privilege men and reinforce their position in the economy, often at the expense of women. What is particularly interesting about this case is that these patterns were reinforced by development nongovernmental organizations (NGOs) that promoted cash crop skills training and technological advancements for men. In essence, existing assumptions about technology and masculinity determined expectations about who should be responsible for techno-solutions to climate change (Bhattarai, Beilin, and Ford 2015). Some NGOs choose to invest their resources in men because they assume that they will be the ones in charge. These gendered assumptions about environmental tasks are not new. In the mid-1980s, Petra Kelly, one of the founders of the German Green movement, warned that "we don't want an ecological society where men build windmills and women silently listen, bake bread and weave rugs" (quoted in Mellor 1992). Examples like these highlight the necessity of thinking critically about connections between gender and climate change, as well as reflecting on power and climate change more broadly.

The women-as-agents discourse reminds us that climate change work is just that—work. It often mirrors the gendered patterns of employment and recognition in other areas of society. Several participants noted that their experiences in the climate change world was remarkably akin to the gendered patterns they witnessed when they worked in other fields. For instance, while sectors such as the clean energy industry or other tech-based climate solutions are commonly touted as an important part of our low carbon transition, they tend to be dominated by the same kinds of people who have historically had a disproportionate say in our future. This is a byproduct of trying to address climate change within existing structures. If we try to "fix" climate change using the same approaches that led to the problem, the same kinds of people are likely to occupy positions at the top. As Elane, a US-based environmental lawyer, observed,

> The proof we have that our climate is changing is science-based, and that science is male-dominated. There's no question about that. But, when we talk about what our approach to addressing this crisis is—I don't think that the framing for explaining the cause and explaining why we believe climate change is real, has to be the same frame for how are we gonna solve this problem? I do believe that a gender-based approach is a solution-based approach because you can't solve a problem with the mindset you used to create it.

In her view, focusing exclusively on science and using those frames at the expense of people-centered approaches is detrimental. A gender-based approach to climate change necessitates posing different questions and unsettling taken-for-granted assumptions.

WHAT DOES THIS BOOK SHOW US ABOUT GENDER?

This book has described gender as a set of socially constructed expectations about behavior that influence the identity and actions of all actors. Gender works on us as individuals by encouraging us to fit our behavior into categories of "masculine" or "feminine." It works on us when we consciously and unconsciously reject these labels, or when we reinforce them. It works on states, multinational corporations, NGOs, and communities in general by shaping their language, networks, and goals. It is a fluid, slippery concept that is nonetheless powerful while sometimes being difficult to spot or recognize. Feminist scholars have produced decades worth of research exploring the idea and are still finding new ways of asking "What is gender?" (Tidwell and Barclay 2019). This is because gender is complex and messy, things that are extremely frustrating to policymakers. Gender works at multiple levels. For instance, on one level gender influences our understanding of what climate change knowledge is and can be. But at the same time, gender shapes the personal experiences of individuals who lack adaptive capacity in the face of climate change impacts. In these ways, gender is a tricky concept because it morphs from changing, often invisible or unacknowledged gendered concepts to the gendered bodies that we typically view as unchanging.

Part of the challenge with the concept of gender in the environmental sphere is that people often have different goals for their analyses. One goal is problematizing taken-for-granted assumptions and getting us to think critically about what gender is, how it works, whom it benefits, whom it marginalizes, and the implications of all these questions (and more). Another goal is to come up with strategies for removing barriers to participation, reducing vulnerability, and working toward sustainability. These are not mutually exclusive. This book is an example of a project that centers on both of these goals and discusses them as necessarily intertwined. We cannot understand why some people are invited to climate change panels and some are not or why some people are less likely than others to possess climate change adaptative capacity without taking gender into consideration.

The failure to connect these various gender goals might lead to unintended negative consequences (Arora-Jonsson and Sijapati 2018). For example, we have to confront the existence of "gender fatigue," or the frustration people express when their attempts to address gender marginalization have failed, and they lack the interest to keep up these kinds of initiatives (Kelan 2009). Several spheres such as human rights, international security, and environmental politics, have adopted policies to "mainstream gender" or "take gender seriously" yet continue to have gendered forms of marginalization and discrimination. When this is pointed out, there may be confusion and frustration about how to move forward. Unintended side effects of this include gender becoming something organizations or institutions are required to "address" without critical reflection on what it is or how meaningful change might be attained. Gender becomes something that is divorced from their day-to-day operations, as organizations might have a one-off gender training session or delegate this work to a "gender person" (Ferguson 2015). Effectively taking steps to minimize or eliminate gendered marginalization and discrimination is difficult and impossible without critical reflection. This is because gender works in ways that are often invisible (Humbert, Kelan, and van den Brink 2019). What happens when we get people to talk about gender, but they do not really care about it? It becomes a box to check. Given the number of gender initiatives instituted in the academic, policy, and nonprofit realms, we seem to understand that it is "important" across these spaces. However, this is a long way from saying we always achieve gender equity or root out the pernicious ways that gender norms sustain patterns of marginalization within these spaces. This book helps us bridge this gap between the messy world of asking questions and critically reflecting on gender and addressing practical problems around gender. This includes questions concerning viable strategies for how to have women's agency and expertise recognized and valued, while also reducing women's physical vulnerability to climate disasters or food insecurity.

For most of my interviewees, gender came down to descriptions of the roles, motivations, and characteristics of women. While all my questions were worded to ask about *gender* and climate change specifically, the vast majority of responses focused on cisgendered women. The ideas of masculinity and men showed up in many interviews, but typically as a contrast to the position of women. Additionally, most responses described gender as a binary (i.e., what fell on one side was excluded from the other side). There was only one reference to transgender women, for instance. This relatively narrow representation of gender is not very surprising, given the way we tend to debate the concept (or not) in common discourse. Many participants mentioned that they often discuss gender dynamics in their workplace, but not typically as part of their routine duties. There were exceptions to this, as some participants commented that their employer had grappled with gender across the organization. However, many more indicated they had not done this work.

The interviewees reflected on gender through both their personal experiences and broad understandings of "women" around the world. In terms of their personal experiences of gender, they mentioned their perception of ways that colleagues relate to them as a woman or how they were socialized to be more receptive to some jobs over others. In many cases, however, their discussion of gender centered on general experiences of women, often characterized variously as nurturing, poor, marginalized, holistic in their thinking, underrepresented in politics, among many others. As argued in chapter 6, these rather simplistic portrayals of women are problematic, particularly as participants often used this broad view of gender to describe women in "developing countries" or "poorer countries" (i.e., the global South). Most representations of women from the global South portrayed them as a homogeneous group that is (1) uniquely susceptible to climate change effects, (2) made up of rural resource users with specific forms of environmental knowledge, and (3) responsible for care work in families. This portrayal of women from the global South illustrates how the various women and climate change discourses overlap. Interviewees tended to use the women-as-vulnerable, women-as-knowledgeable, and women-as-caregivers discourses to describe their position. While this representation of women from the "developing world" undoubtedly describes the lived experiences of many women, it is also far from representing the experiences of all women. Some interviewees were drawing on their research with specific, often rural, communities in the global South. However, most participants were referring to an ideal type that is all too familiar in the global North. This tendency to draw on essentialized portrayals is important because the interviewees are from states that tend to have an oversized role in shaping global discourse about climate change, as well as about gender. The global North has a disproportionately large presence in the academy as well as in global policymaking. If these are commonly used discourses in these states, then this influences climate change debates in important ways.

When gender gets "on the agenda," it is often those in the global North who have determined what it looks like. Moreover, gender—both as a concept and as a policy tool—has a long history of being associated with debate and practice in the global North. María Lugones (2007, 2010), for instance, posits that the colonial project forced particular gendered assumptions and codes of conduct on the colonized. Within the colonial project, the dichotomous hierarchy between humans and nonhuman species became inseparable from the dichotomous hierarchy between men and women. These distinctions determined who was human, and what was civilization.

> Only the civilized are men or women. Indigenous peoples of the Americas and enslaved Africans were classified as not human in species—as animals, uncontrollably sexual and wild. The European, bourgeois, colonial, modern man became a subject/agent, fit for rule, for public life and ruling, a being of civilization, heterosexual, Christian, a being of mind and reason. The European bourgeois woman was not understood as his complement, but as someone who reproduced race and capital through her sexual purity, passivity, and being home-bound in the service of the white, European, bourgeois man. The imposition of these dichotomous hierarchies became woven into the historicity of relations, including intimate relations. (Lugones 2010, 742)

Colonization necessitated transformation, including transformation of identity. Christianity was one powerful tool used for these purposes. It provided the normative rationale connecting gender and civilization and "became intent on erasing community, ecological practices, knowledge of planting, of weaving, of the cosmos" along with reproductive and sexual practices (Lugones 2010). While the formal process of colonization has ended, the meaning and understanding it enforced have lasting implications. These can include simplistic and paternalistic categorizations of people and practices, including flattening the experiences of women from the global South into a singular picture of their lack of agency or capacity.

At the same time, more recent debates about putting gender onto the global agenda requires asking who is primarily taking part in those debates and what does gender look like once there? As discussed in several chapters, institutions such as environmental organizations and the academy have considerable and growing participation by women, but their leadership continues to be male dominated. Likewise, white women are more likely to be represented in these organizations than women of color, particularly in top positions (American Council on Education 2017; Matthew 2016; Taylor 2014, 2018). This might go some way towards explaining the dominance of victimhood portrayals of women from the global South. While historic patterns of marginalization, exclusion, and discrimination have resulted in many women from the global South being acutely susceptible to climate change impacts, only focusing one these communities both erases the experiences of those who do not fit this picture and risks diminishing the agency and personhood of the women who do.

As discussed in each of the book's substantive chapters, looking across the discourses allows us to ask an essential feminist question: Where are the women? Cynthia Enloe (1990) challenged us to pose this question to understand both the presence and absence of women in multiple spaces, as well as the implications of these positions for the women themselves along with the structures they exist within. Taken together, the discourses call to mind multiple forms of women's agency as well as marginalization. Women in the global South were understood to be present in homes, family gardens, and other rural community spaces. They were largely absent from places that we associate with global climate change mitigation. By contrast, women in the global North were regarded as present not only in homes and communities, but also in lecture halls, labs, and conference rooms, where they are active contributors, albeit sometimes undervalued or underestimated.

By pointing out the problematic nature of homogenizing victimization narratives I am not criticizing the participants who utilized these kinds of storylines. I also find myself using them in my classes because students can quickly relate to them. What I am arguing is that it is necessary for us to think through the reasons for the continued use of these simplistic stories. Overall, the discourses used in this book clearly demonstrate that women are already playing numerous roles related to climate change, but that these roles often break down into an "us versus them" pattern.

A lack of critical engagement about gender and climate change connections within climate change spaces in general is apparent from the interviews I conducted. The majority mentioned that they had never really thought much about these connections before I contacted them about the project. And these are people who think a great deal about climate change most days of their lives. Some offered their thoughts as to why that might be so. They noted that it could be in part due to the way they were trained to think about climate change, as well as the mission of the organizations they work for. In the words of atmospheric scientist Victoria,

> I think that's interesting I never really thought about [climate change] in that way [in connection with gender] at all. I think because I've almost been trained when you do scientific work it's always about the data. The data says this; the data says that. It actually wasn't until I went to this workshop where there were people who were truly experiencing it. I mean I get there's a people aspect but

> that's not really what I'm doing, so it's almost like separating the human impact portion of the actual work versus the science that you do. So, I just look at the numbers, you know? But when I met those people it made it feel more real to me. I guess I will point out that I added a chapter to my dissertation or at least a couple of pages in my dissertation that focused on that people perspective which I probably would have never done before. So, I guess that's what made it more real.

Victoria argued that interacting with people who have a different perspective on climate change (i.e., people experiencing climate change effects now) caused her to think differently about it, and even change her research project. I regard this as a good example of how we might observe a change of perspective across multiple communities if we can develop discourses that center the experiences and challenges of women in climate change.

Victoria also attributed her lack of consideration of gender and climate change to her training as an atmospheric scientist. Some might argue that it is too much to require that people should know about all aspects of climate change, including gender. Why should atmospheric scientists have to know about gender and climate change when people outside their professions would not be expected to have their depth of knowledge about climate science? I understand where this argument comes from. However, I would still counter by positing that what I am asking for here is simply a reflection on the fact that the way we frame climate change matters. Currently, our lack of attention to gender and climate change connections means that these interviewees along with scores of others working in these spaces might be more likely to draw on essentialized discourses because that is what they have at hand. The discourses result from us not thinking systematically about gender. When pressed to do so, we have to rely on existing understandings of gender, which tend to be built on problematic, essentialized, and limiting ideas about what gender looks like (MacGregor 2017b). Our lack of deliberation about gender is matched by our lack of reflection about race, class, and other intersectional elements of environmental experience in most dominant spaces of the global North.

All of this is to say that thinking through gender and climate change connections requires walking a tricky path of identifying regularized patterns of behavior while avoiding essentialization. Participants focused mostly on the agency of women in the global North because they were reflecting on their own experiences, or their friends or colleagues' experiences. When they mentioned women who are different from them, such as indigenous women, women in lower socioeconomic classes, and women in the global South, they were thinking of women "out there" rather than "in here." There are important implications to this, including reinforcing power divisions in the international system. This is why specifically incorporating the perspective of people in the global North is an essential complement to all of the work that focuses on people in the global South. We need to understand how we might be drawing on essentializing discourses in climate change debates, and what steps we might take to avoid doing so.

THINKING ABOUT GENDER AND CLIMATE CHANGE INTO THE FUTURE

When I was thinking about who might read this book, several communities came to mind, including people working in climate change spaces and academics working on questions of both climate change and gender justice. The book is premised on a normative commitment to both sustainability and justice. Thomas Princen (2002, 35) has argued that "*sustainability* is not the status quo environment but ecological integrity. [Its] orientation is long term, even very long term, that is, over many generations of key species, including humans. The scale is determined in the first instance by biophysical processes. From this view, human and natural systems may be separate, but the focus is on the intersection of the two systems. Perceived crises demand alternative forms of social organization, ones that make transformational, not marginal, change." This view of sustainability goes beyond a limited view of humanity's ability to continue to exploit nature for our gain. Rather, it captures necessary shifts in our relationships with ecosystems and with each other. At the same time, a basic understanding of gender justice involves seeking to achieve equality of experiences and opportunity. Alison Jaggar (2014, 10) explains that scholars concerned with global justice trace "the ways in which contemporary transnational institutions and recent global policies, most of them facially gender-neutral, have had systemically disparate and often burdensome consequences for specific groups of women in both the global North and the global South." Elsewhere I have argued that achieving gender justice requires reflecting on the sources of gendered disparities and vulnerability and offering fair ways to remove these (Detraz 2017b). In climate change debates, it involves considering the sources of climate change as well as the experiences of climate change, as well as thinking through any benefits to climate change action, and coming up with potential ways to achieve fair distributions of these benefits. While this is an admittedly limited idea of justice as largely focused on equity, I feel that this is a good starting point for larger reflections on gender and environmental issues since we do not even meet a very basic standard of equity in many areas of climate politics (Buck, Gammon, and Preston 2014; Buckingham 2020; Detraz 2017b).

Therefore, this conclusion reflects on how we might rethink each of the four women and climate change discourses discussed in the previous chapters with the goals of sustainability and justice in mind. I suggest ways that academics, policymakers, nonprofit workers, and others can learn from and build on the discourses by thinking about how best to wield these forms of representation in order to achieve sustainability and gender justice goals. This is akin to developing strategic frames, but it goes beyond this. Frames are "'schemata of interpretation' that enable individuals 'to locate, perceive, identify, and label' occurrences within their life space and the world at large" (Benford and Snow 2000, 614). Personal understanding is as an essential component of frames, and strategic framing has been essential for different types of social movements (Jinnah 2011).

What I am calling for are not *just* strategic frames, but rather a reorientation of our perspectives on women and climate change. The discursive shifts that I identify below follow from this reorientation and focus in on specific roles that connect women's lives to climate change challenges. It is my hope that these discourses can also help bridge multiple communities who have compatible goals of sustainability and justice. Despite the fact that they often have compatible goals, global environmental politics (GEP) scholars and feminist international relations scholars often do not talk to each other. They tend to present at different conferences or at least on different panels, publish in different journals, and teach different courses. I hope that this book can play a bridging role for conversations between those who have a normative commitment to sustainability and justice. They are not mutually exclusive, as they reflect our multifaceted experiences of climate change. We know from previous academic work that changes to environmental messaging can result in shifts in attitudes and even increased dedication to environmental protection (Lu and Schuldt 2016; Wolsko, Ariceaga, and Seiden 2016). The nonprofit sector is very aware of this fact, as they consistently work to find effective ways to communicate their climate change priorities. I argue that there are some important shifts that can be made to each of the discourses to make them more inclusive and highlight both agency and obstacles for women in climate change.

Women-as-Vulnerable

The majority of interviewees began their reflections on gender and climate change with a description of women's vulnerability to climate change impacts. While it is accurate that gender and environmental vulnerability have important conceptual and experiential elements (Demetriades and Esplen 2010; Detraz and Peksen 2017; Enarson 2012) it is crucial that we are intentional in our descriptions of vulnerability. We must purposely decouple vulnerability from simplistic descriptions of victimhood. Instead, we must make vulnerability explicitly about marginalization and how we can overcome it. For instance, gender norms create expectations about who will perform specific tasks. The tasks typically deemed "women's work" are less valued (i.e., economically compensated or socially praised). There are typically fewer resources available to support gendered tasks such as caring for families when they become more difficult because of the effects of climate change. Women's adaptive capacity suffers because of the larger social structures that determine what gets prioritized in times of stress or crisis (Denton 2002). It is essential that descriptions of the unique challenges that women face are situated in larger discussions of marginalization and exclusion. As noted in chapter 3, this requires critical engagement with how gender works in a society, but also necessitates making explicit connections between gender, race, class, dis(ability), and place. Though many participants did make connections across forms of marginalization, this must be the central characteristic of a women and climate vulnerability discourse so that we avoid characterizing women as victims.

One way of doing this is to stress that climate change is a phenomenon that is currently impacting the global North as well as the global South. Of course, we need to acknowledge that some communities currently feel the impacts of climate change more acutely than others. But we are past the point of treating climate change as a problem of the global South. Thinking through ways that vulnerability manifests differently in the global North and global South allows for more critical engagement with the forces that influence how we experience climate change, including gender, but also class, race, dis/ability, and so on. Moreover, it facilitates reflection on how we can effectively increase adaptive capacity and ensure human security and environmental sustainability. This reorientation assists us in determining the heightened levels of vulnerability experienced by many rural women in the global South as part of larger patterns that we can work to change. This would allow us to identify sources of vulnerability while ensuring that we avoid the weaponization of vulnerability or the redistribution of vulnerability. We see an example of this reorientation in the language used by a coalition of climate justice and women's rights organizations that came together in 2019 to discuss the Green New Deal concept being debated in the United States and beyond. The coalition was motivated by the idea that feminist analysis is essential to this project. According to its statement of principles,

> To truly address the root causes, as well as the scope and scale of the climate crisis, the Green New Deal must be cross-cutting in its approach, steadfast in feminist principles, and strive to combat historical oppressions. It must advance a transformative feminist agenda that centers the leadership of women, and acknowledges and addresses the generational impacts of colonization and anti-Black racism. It must end oppression against and be led and articulated by frontline, impacted communities—especially women of color, Black women, Indigenous women, people with disabilities, LGBTQIAP+ people, people from the Global South, migrant and refugee communities, and youth. (Feminist Green New Deal 2019)

The steps necessary for achieving these goals include changing economic, political, and environmental systems to redress economic disparity, providing accountability for US policy, and favoring "regenerative, sustainable, cooperative, and collective models" for action (Feminist Green New Deal 2019). This approach situates climate change vulnerability within the existing structures that worsen our climate crisis as well as our justice crisis and sees our future as requiring us to address both simultaneously. Like Stacy Alaimo's (2009, 26) concept of "insurgent vulnerability" or "a recognition of our material interconnection with the wider environment that impels ethical and political responses," a useful, revised climate vulnerability discourse is one that highlights the fact that all living things are vulnerable and identifies those processes that contribute to this vulnerability.

Women-as-Caregivers

The women-as-caregivers discourse underscores how a widely shared human experience, providing care to others, currently intersects with climate change and will do so into the future. This discourse can help shed light on lived experiences as well as underscore the potential for shifts in human-environment relationships. Several interviewees specifically mentioned how they felt personally connected to climate change through their identity as mothers. Linkages between climate change and parenting, which are typically left out of fields such as environmental politics (Princen 2009), are important topics for reflection. At the very least, these storylines offer insight into a potentially important facet of how some may feel a personal tie to climate change, even without any direct exposure to noticeable climate change impacts.

However, descriptions of caregiving roles should avoid homogenizing women as nurturing mothers and avoid homogenizing caregivers to mean only mothers with children. This simplistic portrayal sets unhelpful expectations about who should provide care by depicting women's (and particularly cisgender women's) care work as natural. Since we tend to view gender narrowly as a binary, what falls on one side is excluded from the other. Therefore, since caregiving is often seen as "natural" for women, it must not be for men or anyone outside of the rigid binary. We must recognize these significant pitfalls of uncritical associations between women and nurturing characteristics (Sandilands 1999). We must also recognize that while care narratives are gendered, they are also raced and classed. For instance, some groups of women can use care narratives more strategically than others. Michelle

Carreon and Valentine Moghadan (2015, 30) note that "the status of women within a given society—shaped by race, ethnicity, class, gender, economic status, religion, and sexuality—affects how women experience motherhood, as well as how they are either empowered by or excluded from certain citizenship rights." The "wholesome mother" image is typically seeped in a set of assumptions about class (i.e., wealthy or middle-class) and race (i.e., white). We must be intentional about using a care discourse in ways that incorporate race, class, gender, and other forms of marginalization.

While maternalist frames have been critiqued as potentially reinforcing problematic gender norms (Douglas and Michaels 2005; Gentry 2009), I argue there is still potential in a caregiving discourse for rethinking our orientation to the environment. This would require a more specific focus on *people* as caregivers rather than women. There is a great deal of feminist thought on this topic to draw from (Bauhardt 2014; Di Chiro 2019; Tronto 2013). Some have argued that we are in a "crisis of care" that needs to be addressed as a component of our climate crisis (Bauhardt 2014). In the words of Giovanna Di Chiro (2019, 307), "To reduce our ecological footprint . . . we need to grow our care footprint. What needs to grow is not more green production and green consumption, but a more caring economy rooted in an earth-friendly caring democracy, one that truly 'cares for communities, for future generations in a finite world, and for nature.'" This notion of care illustrates the close connections between sustainability and justice aims.

Several interviewees mentioned that some major transformations are necessary to prevent further climate change. They focused on economics and politics, but also on how we understand human-environmental connections. They questioned whether a new perspective might allow our thoughts and actions to connect to something outside ourselves. How can we get people to understand themselves as part of a whole and act accordingly? Rebecca, a nonprofit worker in the UK, addressed the question directly:

> The best way . . . is to be an agent in a society that is about compassion, and is about care, and is about gratitude, and is about things that mean that we welcome each other and we reach out to each other in a society where people often feel quite insecure or feel quite alone. . . . And only once we connect as a community do we understand the value of being part of the world—our social

> world. I think we're very good at engaging economically but not very good at environmentally or socially.

Saying that we need to teach people how to connect as a community, Rebecca went on to argue that we currently live in "a famine of compassion." This echoes remarks made by other participants who commented on how gender norms affect whether and how men and women are expected to play care roles in society. Shea, an environmental attorney in the United States, likewise claimed that we need a reorientation in ways that we relate to the environment:

> How do we actually change the way that people think about the environment? How do we make people feel about the environment, the way they feel about their own children? How do we bring that relationship temperament to the ground that they walk on?

Shea then argued that a "nurturing piece" that influences how we see the environment and climate change "is missing." According to Rebecca and Shea, we need to get people to recognize connections between human communities in their everyday lives. This builds on the connections that society already expects of women but reorients our assumptions about what this role looks like and who plays it. This necessitates explicitly linking care to humanity rather than femininity.

Existing scholarly work on countries in the global North reveals that emotional connections to climate change are strong (Norgaard 2011) and often feature concern for children or future generations (Fischer et al. 2012). More specifically, research shows that the emotional connection to objects of care under threat from climate change can be a powerful motivator for supporting strong climate action (Wang et al. 2018). These objects of care can be children or future generations more broadly. But it can also be the planet itself. A revised climate change and care discourse would potentially harness this kind of motivation in order to foster a personal connection to climate change, especially among communities who do not consider themselves to be directly experiencing climate change at present. The existing work by parenthood- and motherhood-themed organizations illustrates the salience this frame has for many people since caregiving is such a widely shared human experience. While this is obviously not a salient perspective for everyone and should therefore not be used exclusively, it might be an effective way for some people to view their own experiences as related to climate change.

Women-as-Knowledgeable

The women-as-knowledgeable discourse offers an important counter-perspective to women's victimhood storylines. The discourse furthers sustainability goals by focusing on multiple forms of climate change knowledge that are necessary for understanding of climate change as well as coming up with effective policies to address it. Its storylines contribute to equity goals by reinforcing the idea that women possess climate change expertise and participate in its dissemination in society as scholars, scientists, citizens, or caregivers. It stresses that women are just as capable as men to contributing to climate change understanding, and in some instances have unique climate change knowledge due to gendered divisions of labor in society or households.

In this way, the discourse acknowledges women's expertise in multiple forms. It confirms that women are already represented in roles associated with climate change knowledge, including as negotiators, researchers, and teachers. At the same time, it illustrates transformative potential in how we conceptualize knowledge in the context of climate change. For instance, some participants mentioned that the people who are perceived as climate change stars are those who dedicate an enormous amount of time to researching climate change and publicly exhibiting their knowledge. In one participant's view, "They dedicate day and night to reading articles and attending meetings. And I think as a culture, we idolize that, the people who know all the stuff are like gods." While we obviously need passionate, dedicated, and knowledgeable people working in climate change spaces, not everyone has the same chance to belong to this group. Chapter 6 explored the theme of using one's position as "superstar" to exclude others. One interviewee used the term "silverbacks" to refer to older white men in her field who are regarded (and view themselves) as the "heavy hitters" who should be included on conference panels. Another participant pointed out that some older white men in her atmospheric science department think of themselves as "the most important people in the department" because of their status in the field. These examples illustrate the potentially problematic nature of seeking to highlight women's position in existing structures of knowledge production. These structures are gendered, raced, and classed in ways that perpetuate a perspective of knowledge that is centered in male-dominated institutions in the global North. Kavita Philip (2020) notes that dominant, northern understandings of knowledge have a long history of taking information and lived experiences from other places and representing them as their own accomplishment and expertise.

> Medical pharmacopeias were written with the help of non-Western peoples; for example, Indigenous populations in European imperial contexts provided information about the use of plants. Astronomical and mathematical knowledge traveled from South Asia and the Middle East to Europe, whose "Renaissance" would not have been possible without it. So, in a sense what we call "the scientific method" and "objective" language is itself [*sic*] the product of an imagined autonomy of educated, Western men, who took the knowledge of the world and embedded it in their own claims to mastery. In the process they told a story about the legitimate subject of knowledge, suggesting that only certain kinds of people could be knowers, thinkers, intellectuals. Women, Indigenous people, and enslaved and colonized people lost even more than their individual lives and liberty—they lost, for over two centuries, the right to be considered knowledge producers. We still live in institutions that were created based on this assumption. (Causevic et al. 2020, 25)

The academy is notorious for funneling information through "proper" channels and rejecting that which is not "rigorous." For many people doing this evaluating, a lack of rigor sometimes just means unfamiliar. Again, this does not suggest that we do not need accurate, scientifically grounded information about climate change. We need women doing the "fabulous science" that one participant mentioned in chapter 5. This is important, and their accomplishments should not be underestimated. But we also need other information in order to achieve sustainable and just climate policies. It is therefore important to frame the idea of women as knowledgeable with an expansive conceptualization of what constitutes expertise and how it is generated. This would require drawing on examples both inside and outside dominant academic as well as social structures.

We leave out valuable forms of knowledge if we limit ourselves to a narrow definition. I argue that inclusion must encompass expanding not only on the range of perspectives that are physically present in climate change spaces, but also on conceptualizations of knowledge and expertise more broadly, as well as explicitly "centering knowledges from the margins" (Causevic et al. 2020). In climate change spaces, the margins include the global South, as knowledge about climate change (along with many other subjects) tends to be created in the global North about the global South (Blicharska et al. 2017). The margins also entail those spaces outside of our dominant frame of reference for expertise.

Along these lines, several interviewees brought up examples of women's (and men's) practical knowledge about climate change effects. This avoids painting women as knowledgeable only if they have a specific set of credentials. However, calling for an expansion of our view of knowledge does not mean treating women's knowledge as simply an input into climate change understanding or engaging in "epistemic objectification." Doing so reinforces problematic patterns of exploitation (Tuvel 2015). Rather, it means recognizing that "knowledge has multiple forms—both formal and lived. Knowledge is a process, not a product" (Causevic et al. 2020). This reorientation concerns those involved in the process as participants.

Interviewees frequently stressed the importance of knowledge that emerges from lived experiences. One example is women's early awareness of climate change impacts because of difficulties in subsistence agriculture or rising food prices at markets. While calling attention to alternative ways of knowing about climate change is in itself positive, we should still avoid automatically associating local knowledge gained from gendered household tasks or resource use with women, and particularly women of the global South. This is in part because this knowledge can be exploited without them being incorporated as full partners in climate action (Dey, Singh, and Gupta 2018; Dove 2006; Federici 2009; Sapra 2009). There is a delicate balance between accurately portraying lived experiences and avoiding representing these as the only possibilities. We must strive to accept the climate change knowledge that comes from gendered tasks while recognizing that these roles are, in fact, gendered.

Women-as-Agents

The women-as-agents discourse highlights women's active participation in multiple areas related to the causes of climate change, along with attempts at climate change mitigation and adaptation. Many of the storylines in the discourse focus specifically on women's activity in the environmental movement, particularly due to the significant presence of women in environmental organizations. It sheds light on different ways women participate in climate change action and possible consequences of this participation. The discourse is also useful for reminding us of where there is more to be done to achieve increased gender equity in climate change work by highlighting where expertise is undermined and action is blocked or undervalued. It contributes to sustainability aims by calling attention to multiple sites of climate change action that can be enhanced in the future. It furthers justice goals by shining a light on the many women across the globe who are already climate change actors as well as the gendered resistance these women sometimes face.

Many participants used the women-as-agents discourse to describe women's many contributions to climate work. The discourse is therefore beneficial for countering an understanding of women as passive victims of climate change. However, we must be wary of using stereotypical representations of women's environmental action that reinforce expectations that environmentalism is an inherently feminine sphere. Some scholarly work raises the concern that these associations can serve to discourage some men from environmental action (Brough et al. 2016; Swim, Gillis, and Hamaty 2020). I argue that it is also problematic because it offers homogenizing assumptions about women. While it may be the case that some women are more likely to be drawn to environmental work because of gender socialization, this socialization piece of the story needs to be made clear. We must avoid depicting women's climate action as a byproduct of their inherently nurturing or giving "nature." There are gendered, raced, and placed implications of these associations, as women from the global North are often the ones assumed to be environmentally virtuous (Arora-Jonsson 2011). These depictions of certain environmental actions by women as "natural" mask the struggles that some environmental activists have encountered in getting their concerns taken seriously. For instance, environmental justice–oriented organizations, many of which are led or heavily staffed by women of color, have historically been less funded and less connected to centers of power than more mainstream environmental organizations (Bullard and Smith 2005). Additionally, discussing women's environmental agency necessitates considering the potential risks or dangers that people can face when they engage in environmental protection activities—including criminalization or violence (Glazebrook and Opoku 2018; Tran et al. 2020). Agency is complicated by patterns of marginalization, and a women-as-agents discourse should reflect this.

The women-as-consumers storyline in the women-as-agents discourse is particularly useful for demonstrating that agency is complicated, because the storyline forces us to reflect on the economic processes that are inextricably tied to environmental change and how these intersect with gender (Dauvergne 2008, 2010). It also draws attention to women's participation in environmental damage because of their role as consumers. While consumption patterns are gendered, many women, particularly many women in the global North, are active participants in our current consumer culture. Hence, the women-as-consumers storyline calls attention to women's participation in multiple facets of environmental change, some good and some bad. It also allows for the consideration of consumption itself as an important source of environmental damage, one particularly associated with historic patterns of economic activity in the global North. This is important since the issue of population growth in the global South has historically attracted much more attention in environmental policymaking (Hartmann 2010).

Likewise, it is important to address the forms of exclusion that were identified by participants, as well as some forms that were not brought up in the interviews. Most interviewees focused on women's exclusion from or underestimation in multiple professional spheres. While there was some discussion of other forms of exclusion—race, ethnicity, class, and place in particular—these should all be central to a quest for climate solutions that are sustainable and just. One way of addressing these forms of exclusion is to recognize their intersectional nature. Some participants did so, for example, by reflecting on trends such as women of color being underrepresented in climate science or indigenous women in particular being absent from climate negotiations.

This kind of intersectional analysis ensures that calls for greater inclusion do not result in greater inclusion of only white women from the global North.

These transformations to achieve greater inclusion across multiple communities of marginalized peoples are part of the discursive shifts and social changes that will be essential for effective and just climate change approaches. Kate O'Neill, Jörg Balsiger, and Stacy VanDeveer (2004,152) define social change as "a process by which the interaction between agents and structures creates new possibilities for collective action by changing norms and institutions, as well as the evolution of existing and emergent actors (and their interactions) who are both enabled and constrained in the pursuit of their goals." Some participants, particularly from the scientific community, pointed out that their fields are witnessing a transformation as more women are rising through the ranks. Having more women in fields like this offers some transformative potential. Alternative voices and lived experiences will hopefully contribute to changing norms and possibly changing the institutions themselves. Although there is always the possibility that any given institution will remain the same while simply having different people in them, many interviewees were optimistic about the effects the increased levels of gender equity. Yet, they still argued for important changes to take place, including having more women, people of color, and others in high-level or leadership positions in the areas that significantly contribute to climate change action. Changing the discourses we use to describe and understand gender and climate change will be one major way to contribute to this movement. Luckily, we have decades' worth of feminist environmental scholarship to draw on in this quest. Sherilyn MacGregor (2006, 6), for instance, has argued in favor of a project of "feminist ecological citizenship," arguing that bringing the concept of citizenship into environmental debates "provides an inclusive space for the public performances of political subjectivity that destabilize and resist dominant ideologies of gender." This idea of feminist ecological citizenship represents an important type of transformational environmental agency that seeks not only to "add women," but to fundamentally alter our way of thinking about power structures within society. Likewise, Anna Kaijser and Annica Kronsell (2014, 428) remind us of the necessity of unraveling the multiple forms of power that shape not only our agency but also our understanding of global challenges such as climate change.

> Intersections of power structure political, economic, and social institutions on all levels. It is therefore necessary, in an intersectional analysis, not only to look for the adverse impacts of climate change on "vulnerable" groups, but also to shed light on and problematise norms and underlying assumptions that are naturalised and regarded as common sense, but build on and reinforce social categorisations and structures of power, not least through institutional practices. We need to turn our gaze towards economic elites and the Western countries. . . . Using intersectionality in the study of climate issues makes it possible to reach a more complete and accurate understanding of the social and political conditions for climate governance.

This view is essential for reevaluating each of the four women and climate change discourses outlined in this book. While it is undoubtedly important to challenge essentializing discourses that paint the world's women as victims of climate change, each of the discourses has space for revision in order to provide transformation necessary to enhance sustainability and justice.

ONWARD

I started this project because I was curious about the various ways gender and climate change intersect. I wanted to uncover how people working in climate change viewed these connections. Over the course of seventy-six interviews (and twenty survey responses), I found myself feeling hopeful. This hope stems from the knowledge that there are dedicated people working on climate change daily. It comes from their confidence that although we have a huge amount of work to do on both climate change and gender equity, there are bright spots and points of progress. This project's discursive map and suggestions for discursive shifts should therefore be helpful for them and others who work on climate change. Additionally, the discourses uncovered in this book are useful for policymakers, the media, scholars, and everyday people. This is particularly true for youth activists who have a significant stake in what direction our future will take. Using discourses of women and climate change that highlight agency and ways to address historical marginalization are essential for the next generation to break out of existing patterns of casting southern women as climate victims. We have seen that environmental organizations have at times struggled to break from their patriarchal past, but as discussed in chapter 5, some patterns of masculinity being rewarded appear in youth climate movements as well (Curnow and Chan 2016).[2] As the youth climate movement continues to struggle with ageist representation by the media and policymakers (Bergmann and Ossewaarde 2020; Kimball 2019; Mayes and Hartup 2021), a deeper understanding of gender and climate change is relevant for understanding what future paths of justice and sustainability might look like.

A central purpose of this book is to alter people's perceptions so that when they hear the phrase "gender and climate change," they do not see only images connected to women's physical climate change vulnerability. I want them to envision my interviewees and all of the important climate change work they are doing, along with all of the women across the global North and global South with whom I did not have the opportunity to talk. Women are connected to climate change through multiple roles and experiences. Ignoring this reduces women's agency and renders their contributions less visible.

The four discourses—women as vulnerable, caregivers, knowledgeable, and agents—provide ways of understanding women and climate change connections that are grounded in lived experiences. In addition, the discourses speak to broader trends and debates about climate change. Though they sometimes replicate stereotypical depictions of women as nurturing or marginalized, they challenge essentializing storylines by highlighting women's expertise in climate change or contribution to the causes of climate change through consumption. As discourses they represent ways of knowing and understanding climate change for the interviewees. They are the stories they, and often we, tell about climate change drawn from lived experiences, socialization, and available information. The women who spoke with me presented a picture of the lives of "women." Sometimes this included critical evaluations of gender and power or problematic essentialization about women in "poor countries" or proud declarations about the many contributions women have made to the field. They are stories, which are created and alterable. I argue that there are some important alterations necessary to each discourse that would allow them to function as frames of representation to highlight the multifaceted roles of women in climate change. This is vital because representation matters. If we see or hear women discussed as victims, then we will have a hard time imagining them in global negotiation spaces, or atmospheric science conferences, or in board meetings of climate change organizations. These are spaces where women routinely exist and contribute. Most especially, representations of women from the global South matter, as they tend to be the ones associated with vulnerability and victimhood. It is crucial to unpack assumptions about agency and use discourses that emphasize women's participation as well as obstacles to their inclusion.

In sum, gender needs to be a central component of how we think about and act on climate change. We need to recognize that gender shapes climate change at multiple levels. This ranges from influencing people's individual experiences of climate change impacts (Dankelman 2010; Detraz and Peksen 2017) all the way up to shaping the mitigation and adaptation policies we adopt (Dankelman 2002; Demetriades and Esplen 2010). We need critical engagement with the roles and responsibilities assigned to "women" because many people are still surprised to see the variety of roles that women currently play in climate spaces. This indicates that we need more critical deliberation of gender and climate change connections across multiple sectors working on climate change. That has been the central goal of this project: to facilitate an evaluation of where women are currently playing roles, what these look like, and what the future brings.

Appendix: Participant Information

Participant	Location	Length of Time in Climate Change	Current Climate Change Work	Method of Interview and Contact*
Abigail	US	10 years	Works for a clean energy nonprofit and is responsible for multiple facets of advocacy work	Video; secondary
Aida	US	4 years	Works at a university studying climate change impacts on storm surge	Video; secondary
Alice	UK	2.5 years	Works for a climate nonprofit as a knowledge manager	In person; primary
Allison	US	15 years	Studies environmental governance and teaches courses on climate change at a university	Phone; primary
Annica	US	23 years	Social scientist who researches and teaches climate change	Video; secondary
Annise	US	2.5 years	Works for an environmental nonprofit as an events manager and part of the development team	In person; primary
April	Germany	4–5 years	Works for a climate nonprofit in the education sector	In person; primary
Ashley	UK	13 years	Researches climate change ethics and justice at a university	In person; primary
Aubrey	Germany	3–6 years	Works for an organization that focuses on climate change innovation and runs education programs	Video; secondary

(continued)

Participant	Location	Length of Time in Climate Change	Current Climate Change Work	Method of Interview and Contact*
Belina	UK	8 years	Works for a climate nonprofit and does public engagement and social media tasks as well as conducting research on population growth	Video; primary
Brenda	US	3.5 years	Runs the climate and clean energy program for an environmental nonprofit	In person; primary
Brittany	UK	6 years	Works for an environmental nonprofit on the communications team	In person; primary
Brooke	US	20 years	Studies climate governance at a university	In person; primary
Candice	US	25 years	Researches ice in the climate system at a university	Video; secondary
Carin	UK	7 years	Studies social perceptions and implications of weather and climate change at a university	In person; primary
Carolina	UK	3 years	Works for an energy advocacy group on communications projects about the domestic energy system	In person; primary
Claire	UK	8 years	Works at an environmental nonprofit on sustainable living initiatives	In person; primary
Constance	US	3 years	Researches volcanic eruptions and climate	Video; secondary
Darcy	US	3 years	Researches and teaches gender and climate change at a university	In person; primary
Deb	UK	2.5 years	Researches severe weather effects on lifestyles and livelihoods	Video; primary
Elane	US	10 years	Works as a legal advocate for an environmental organization	Video; primary
Eva	Germany	8 years	Works for an environmental nonprofit on climate education initiatives	In person; primary
Farhana	UK	2 years	Works for a faith-based nonprofit on climate change projects	In person; primary

Participant	Location	Length of Time in Climate Change	Current Climate Change Work	Method of Interview and Contact*
Fiona	US	5 years	Researches interactions between animals and the environment	In person; primary
Ginnie	UK	8 years	Works for an environmental nonprofit on sustainable living campaigns	Phone; primary
Glenda	UK	24 years	Works as an attorney for an environmental nonprofit	In person; primary
Gwen	US	5 years	Works for an environmental nonprofit coordinating its federal and state advocacy work	Phone; secondary
Haley	US	1.5 years	Works for a climate change nonprofit doing communication, policy, and organizing	Phone; secondary
Heather	US	9 years	Researches how climate change will affect plants and animals at a university	In person; secondary
Heidi	US	half a year	Works for an environmental nonprofit on solar projects	In person; secondary
Hildi	UK	3 years	Works for an environmental nonprofit on sustainability and energy projects	In person; primary
Ingrid	Germany	3 years	Works for a climate nonprofit on programs related to reducing emissions from international shipping	Video; primary
Ivy	Canada	6 years	Researches and teaches gender and climate change	Video; primary
Jane	US	2 years	Works for an environmental nonprofit on initiatives focused on the transition away from fossil fuel use	Video; primary
Janice	US	7 years	Works as the deputy director of a climate nonprofit	In person; primary
Jasmine	US	10+ years	Studies climate change impacts on plants at a university	Video; secondary
Jessie	US	12 years	Works for an environmental nonprofit doing outreach to policymakers and community members	Phone; primary

(continued)

Participant	Location	Length of Time in Climate Change	Current Climate Change Work	Method of Interview and Contact*
Joni	US	5 years	Researches climate change in urban ecosystems	Phone; secondary
Julia	UK	10 years	Researches forestry mitigation initiatives at a university	In person; primary
June	US	2 years	Works for an environmental nonprofit as a communications specialist	In person; primary
Kate	US	2.5 years	Works for an environmental nonprofit as a grassroots organizer	Phone; primary
Kathryn	US	9 years	Works for a climate nonprofit as a communications manager	Phone; secondary
Kimberly	Germany	2 months	Interns at an environmental organization and researches sustainability incentives in cities	In person; secondary
Kit	UK	20 years	Studies community-based environmental management initiatives at a university	Video; primary
Kristy	US	18 years	Researches water supply and climate change as well as doing outreach and public engagement work	In person; secondary
Kylie	US	25 years	Researches perceptions of climate change and teaches courses on climate change	Phone; primary
Lily	UK	1 year	Works for an environmental nonprofit on community engagement and education in waste reduction	In person; secondary
Linda	US	5 years	Works for an environmental nonprofit focused on children's health	Phone; primary
Lydia	US	5 years	Works for a solar company	Video; secondary
Lyra	Germany	3.5 years	Works for an environmental organization coordinating with delegations to international climate negotiations	Video; secondary
Marie	Germany	7 years	Works for an environmental nonprofit on projects that facilitate dialogue between European cities on climate policy instruments	Video; secondary

Participant	Location	Length of Time in Climate Change	Current Climate Change Work	Method of Interview and Contact*
Mary	UK	7–8 years	Researches climate governance and teaches classes on climate change at a university	Video; secondary
Meg	US	4 years	Works for an environmental nonprofit as a field organizer	In person; primary
Molly	US	20 years	Researches nonstate actors in the climate regime as well as teaches classes on climate change at a university	Phone, primary
Nicola	US	3 years	Works for an environmental nonprofit in communications and outreach	In person; primary
Nora	UK	10 years	Works at a university managing a research program on climate change and energy	In person; primary
Paige	UK	6 years	Co-directs an environmental nonprofit	In person; primary
Quinn	US	5+ years	Works for a faith-based organization to spread awareness of climate change issues	Phone; primary
Rachel	US	30 years	Runs a research group on climate change at a university	In person; primary
Rebecca	UK	10 years	Works for an environmental nonprofit on waste prevention initiatives	In person; primary
Sarah	US	30 years	Directs an interdisciplinary research program on climate change at a university	Phone; secondary
Selma	Germany	16 years	Works for an environmental nonprofit on multilevel climate governance projects	In person; primary
Sharon	US	10 years	Works for a clean energy nonprofit on coal plant retirement initiatives	In person; primary
Shea	US	3 years	Works for a consumer advocacy organization on solar projects	Video; primary
Sina	US	9 years	Works for a faith-based nonprofit to encourage climate change education for religious leaders	Video; secondary

(continued)

Participant	Location	Length of Time in Climate Change	Current Climate Change Work	Method of Interview and Contact*
Swati	US	20 years	Works for an environmental nonprofit in fundraising	In person; secondary
Talia	Germany	>6 months	Works for an environmental nonprofit on carbon emissions reduction projects	In person; secondary
Vanessa	UK	1 year	Works as for an environmental nonprofit on transportation and sustainability initiatives	Phone; primary
Vicky	UK	5 years	Works for an environmental nonprofit on food waste and sustainability initiatives	Video; secondary
Victoria	US	10 years	Researches clouds and climate change as well as performs education outreach on climate change in K–12 schools	Video; secondary
Yvette	UK	22 years	Works for an environmental nonprofit in environmental education	Phone; primary

* For the "Method of Interview and Contact" column, the method of interview means how the interview was conducted—either in person, by video (including Zoom, Skype, or Google Hangouts), or by phone. The method of contact indicates whether the participant was contacted directly (primary) or was identified using the snowball method (secondary).

Notes

CHAPTER 1

1. According to the IPCC (2014a, 6), "People who are socially, economically, culturally, politically, institutionally, or otherwise marginalized are especially vulnerable to climate change and also to some adaptation and mitigation responses. This heightened vulnerability is rarely due to a single cause. Rather, it is the product of intersecting social processes that result in inequalities in socioeconomic status and income, as well as in exposure. Such social processes include, for example, discrimination on the basis of gender, class, ethnicity, age, and (dis)ability."
2. Oceania was the region that saw the largest increase in media coverage of climate change between 2004 and 2010. North America was second, and Europe was third (Boykoff et al. 2020).
3. Levin et al. (2012) draw on the earlier concept of "wicked problems" from Rittel and Webber (1973).
4. It is interesting to note that while climate change research dominates global environmental politics, it is vastly understudied in the larger field of political science. See, for example, Javeline (2014) and Keohane (2015) for why this is problematic.
5. For other work in this vein, see Pettenger (2007).
6. Ecofeminism is one of the most well-known variations of feminist environmental work.

CHAPTER 2

1. Tickner (1997, 613) outlined three types of misunderstandings that often occur between feminist scholars and others in the field: "first, misunderstandings about the meaning of gender as manifested in the more personal reactions; second, the different realities or ontologies that feminists and nonfeminists see when they write about international politics, evident in comments that feminist scholars are not engaging the subject matter of IR; third, the epistemological divides that underlie questions as to whether feminists are doing theory at all."

2. Problem-solving theories, on the other hand, are status-quo oriented and largely take the prevailing structures, institutions, and power relationships as the given framework for action.
3. O'Neill et al. (2013) have argued that methods are often underspecified in key GEP research areas.
4. This is echoed by Haas and Mitchell (2013, 73) when they say that "although the causes and effects of global environmental problems tend to be multidisciplinary and interdisciplinary, modern scholars too often are disciplinary. The complexity of environmental issues—in terms of the number of and interactions among variables, the length of causal chains, and the extent of interactions across time, space, and scale—requires insights from multiple disciplines to capture accurately the extensive and multiple understandings of their causes, causal mechanisms, and effects."
5. The appendix contains information about the participants, including their location, length in climate change work, and type of current climate work.
6. I contacted nine environmental justice organizations and was unable to arrange interviews with any of their members. I also contacted two women and environment organizations and was unable to set up any interviews with their members.
7. The appendix indicates which participants were originally contacted and which were recommended by one of the interviewees.
8. The transcriptions were analyzed verbatim. The direct quotations that I include in the manuscript have been lightly edited for clarity. This entailed removing words or sounds such as "like" and "um."
9. The surveys yielded a slightly different distribution of the discourses as the interviews. The women-as-vulnerable discourse was the most frequently used in the surveys (used twenty-three times). This was followed by the women-as-knowledgeable (eighteen times), then the women-as-agents (fourteen times) discourses. The women-as-caregivers discourse was only used once. However, the storylines of the discourses from the surveys were consistent with those from the interviews.
10. The women-as-agents discourse was the most frequently used in the interviews (used four hundred six times). The women-as-caregivers discourse was used two hundred fourteen times. The women-as-vulnerable discourse was used one hundred twenty-two times in the interviews. The women-as-knowledgeable discourse was used seventy-two times.
11. Participants have lived or conducted research in Bangladesh, Congo, Egypt, Gambia, Guatemala, India, Indonesia, Kenya, Nepal, Nigeria, Tanzania, Tunisia, Uganda, and Vietnam.

CHAPTER 3

1. Farhana is also centrally involved in efforts to raise awareness and encourage climate action in Muslim communities.
2. Studies have also focused on specific sectors of society that will be susceptible to the effects of climate change; these include, for example, agriculture (Challinor et al. 2007) and tourism (Scott, Hall, and Gössling 2019).

3. The study focuses largely on economic vulnerability to climate change impacts. While the focus on female-headed versus male-headed households offers a unique approach to gendered differences of experience, it does not address gendered differences within households, something the authors acknowledge (Andersen, Verner, and Wiebelt 2017).
4. Participants working in different climate change sectors—such as academics in the natural and social sciences and those in nonprofits—used the women-as-vulnerable discourse.
5. The women-as-water-collector storyline came up twenty-five times across the interviews, women-as-wood/fuel-collector nine times, women-as-food-provider twenty-three times, and women-as-agricultural-worker seventeen times.
6. More will be said about specific sources of knowledge in chapter 4.
7. Nine participants specifically mentioned women's experiences in natural disasters in their responses.
8. We see this focus on climate change exacerbating patterns of marginalization in international policy documents (IPCC 2014b).
9. I asked whether participants saw any connections between the causes, experiences, and solutions to climate change, among other questions.
10. Early environmentalism in the global North has often been critiqued for its discourse of claiming natural spaces. For instance, within the history of German environmentalism ideas of sustainability and conservation were embedded in the spirit of military readiness and nationalist expansion from the nineteenth century (Markham 2008). In the United States and Canada, "conservation was tied to racist, sexist, and classist notions of wilderness protection in order to serve urban, bourgeois, white men's desire to construct themselves as rugged frontiersmen" (Curnow and Helferty 2018: 148).
11. See Miranda Schreurs (2004) for an assessment of how this series of transformations played out in states such as Japan, Germany, and the United States.
12. The male-dominated nature of mainstream environmental organizations will be expanded on in chapter 5.
13. This frustration with the lack of progress on social justice issues within mainstream environmental organizations will be discussed further in chapter 5.
14. Their study specifically examined the impact of Federal Emergency Management Agency (FEMA) programs.
15. Important exceptions include work like that of Carrigan (2010); Priestley and Hemingway (2007); Wolbring and Leopatra (2012).
16. The IPCC (2014a, 1066–1067) report on impacts, adaptation, and vulnerability does include mentions of disability such as this one: "Wealth, education, ethnicity, religion, gender, age, class/caste, disability, and health status exemplify and contribute to the differential exposure and vulnerability of individuals or societies to climate and non-climate related hazards."
17. Fothergill and Peek (2015, 223) focus their natural disaster study on individuals aged three to eighteen. They explain that "children around the world are considered a social minority

group because they are marginalized in terms of wealth, social status, and political power." For other work on youth and climate change, see Tobin-Gurley et al. (2016).

18. More will be said about this topic in chapter 4.
19. Notable exceptions include discussions of air quality and pollution in cities. See chapter 3 for more about asthma rates and climate change.
20. Participants in this study have lived or conducted research in Bangladesh, Congo, Egypt, Gambia, Guatemala, India, Indonesia, Kenya, Nepal, Nigeria, Tanzania, Tunisia, Uganda, and Vietnam.
21. These mentions of the unique position of "women in the global South" were sometimes coupled with a reflection on the position of marginalized peoples in the global North as well some similarities and differences between them.
22. The Intergovernmental Panel on Climate Change (IPCC) and the United Nations Framework Convention on Climate Change (UNFCCC) have long focused on climate change vulnerability across multiple underrepresented groups (Füssel and Klein 2006; Thomas and Warner 2019).
23. According to Rao and Raj (2019), "There is global consensus that women are integral to climate change dialogue, not just because of their role and dependence on natural resources, but also because of their disproportionate vulnerability to climate change threats. Yet, there is a paucity of data documenting . . . women's roles and engagement in climate change adaptation. We could identify no single standard measure focused on these issues. Global indicators on women and climate change action are lacking."
24. At the same time, the 2019 UN Environment Programme (UNEP) resolution on gender equality in environmental governance is another example of a policy document that overwhelmingly adopts a women-as-vulnerable discourse, with women depicted almost exclusively as marginalized and in need of empowerment.
25. Racial composition of a neighborhood is a strong predictor of the siting of facilities that product environmental damage (Bullard 2005).
26. There are many reported cases of women experiencing violence in relief camps after natural disasters (True 2012).
27. Environmental decision-making often excludes marginalized communities such as indigenous peoples (Agyeman, Bullard, and Evans 2003).
28. This is often conceptualized as a form of "maladaptation." Barnett and O'Neill (2010, 212) explain that "adaptation strategies may increase the vulnerability of other systems, sectors, or groups if they increase emissions of greenhouse gases, disproportionately burden the most vulnerable, have high opportunity costs, reduce incentives to adapt, or set paths that limit the choices available to future generations."

CHAPTER 4

1. Instead, Sherilyn MacGregor (2006, 6) has argued in favor of a project of "feminist ecological citizenship," arguing that bringing the concept of citizenship into environmental debates "provides an inclusive space for the public performances of political subjectivity that destabilize and resist dominant ideologies of gender."
2. According to Cameron Butler (2017, 275) queer ecology scholars "use the word 'queer' as an adjective to describe sexuality and gender, but they also use it as a verb, as in 'to queer,' to engage in the act of queering. *To queer* is to question the categories, definitions, divisions, distinctions, and dualities that exist and that usually go unchallenged within society."
3. The women-as-caregivers discourse was used 214 times across the interviews by both academics and women in the nonprofit sector.
4. See Alscher (2011) and Evertsen and van der Geest (2020) for accounts of women's migration being at least partially tied to environmental factors.
5. As mentioned above, gender is significantly associated with the extent of concern about specific environmental problems (McCright 2010; Mohai 1997; Xiao and McCright 2012).
6. Other groups include older people, people with existing health problems or disabilities, and poor and marginalized communities (Watts 2015).
7. The idea of women providing food or creating an inviting space came up in multiple interviews. Women were identified as teamakers, as providing food for meetings, etc.
8. One interesting contrast to this is the image of men in the environmental movement as performatively caring. Sharon, a US nonprofit worker, brought up the concept of the "mactivist"—a man who, in her words, uses "his caring and love to play around with the women in the movement."
9. Population discourses have also specifically been applied to climate change through concerns about climate-induced scarcities and/or climate refugees (Hartmann 2014).

CHAPTER 5

1. The women-as-knowledgeable discourse was used seventy-two times across the interviews, making it the least frequently used discourse of the four outlined in the book.
2. Guy et al.'s 2014 study of people in Australia found that those who had greater knowledge of climate change causes were more willing to accept that climate change is occurring.
3. Several studies focus on specific states or communities, typically within the global South, to highlight women's environmental knowledge. For instance, Lane and McNaught (2009) focus on the Pacific region, while Agarwal (2009) focuses on India and Nepal.
4. It is interesting to note that Evangelical individuals with lower incomes were also more likely to assess climate change as riskier than individuals with higher incomes. These findings suggest that people who have experienced marginalization in some form or another may be more cautious in their assessment of risk (Smith and Leiserowitz 2013).

5. Multiple of academic studies focus on conservative men in particular. For instance, Krange, Kaltenborn, and Hultman (2019) examine conservative Norwegian men and find that 63 percent of men in this category do not believe in anthropogenic climate change, while only 36 percent of the general population express climate denial.

6. "Drawing from examples in Western Europe and the US, we focus on those whose primacy blinds them to their impacts on society and environment—individuals and constituencies who are enmeshed with fossil-fuel addicted industrialisation and corporatisation, are commonly aligned with climate change denial and whose allegiances are emboldened by traditional socialisations of masculine identities that we refer to as 'industrial/breadwinner masculinities'" (Pulé and Hultman 2019, 86). More will be said about this in chapter 7.

7. According to a twenty-six-country study by the Pew Research Center, climate change is regarded as a substantial global threat in regions across the world. In fact, in thirteen countries climate change was identified as the top global threat (Poushter and Huang 2019).

8. One interesting aspect of the report is the increase in understanding of climate science. Over half (57 percent) of the respondents understand that there is scientific consensus that climate change is occurring. This marks the highest level of assessed scientific certainty since 2008. However, this number drops significantly when respondents are asked about how strong the level of consensus is. Only one in five (20 percent) understand that over 90 percent of the scientific community agrees that human-caused climate change is happening (Leiserowitz et al. 2018).

9. Climate change is seen as the most serious global problem in Sweden (38 percent) and Denmark (29 percent). Less than one in ten respondents in several countries in Southern and Eastern Europe reported this level of concern (European Commission 2017).

10. Participants used the phrases "climate change denial" or "climate skepticism" to refer to the position of believing that climate change is not happening or that humans are not the main drivers of climate change.

11. This theme will be taken up again in chapter 6, which notes how participants censored their discussions of climate change so as not to cause tension or be exposed to ridicule.

12. These connections were also made in some of the survey responses. For example, one respondent said, "Women are often the purveyors of seeds and because of their traditional role as household caregivers they are concerned with getting the most nutrition into their families."

13. This theme of the necessity of women's participation in policymaking will be taken up in chapter 6.

14. Participants used a women-as-climate-change-leaders storyline fourteen times in the interviews.

15. Seven participants mentioned Christiana Figueres, five by name and two by referring to her position as leading negotiations for the Paris Agreement.

16. While there is much disagreement about the promise or shortcomings of the Paris Agreement, environmental politics scholarship has long revealed the difficulty of getting widespread participation in a binding multilateral agreement such as this one (MacGregor 2017a).

17. An exception to this storyline of women's participation in climate change diplomacy/negotiations was Elane's note that Middle Eastern countries are still reluctant to let women play a role in these negotiations. Kristy made a similar comment about the prevalence of women at global climate negotiations being mediated by gender norms in the country.
18. These figures cover the gender composition of constituted bodies established under the UNFCCC, the Kyoto Protocol, and the Paris Agreement (UNFCCC 2019).
19. Women's advocacy on climate change issues is also a frequently mentioned storyline and is discussed in chapter 6.
20. The finding that women are systematically cited less than men holds even when controlling for year of publication, venue of publication, substantive focus, theoretical perspective, methodology, tenure status, and institutional affiliation (Maliniak, Powers, and Walter 2013; Mitchell, Lange, and Brus 2013).
21. Along these lines, some interviewees drew from their experiences in the classroom to respond to my questions about gender and climate change connections. Annica, for instance, claimed that she has seen more interest in environmental topics from her female students than her male students. Darcy said that while she had not really reflected on it at the time, most of the students who reacted negatively to her critiques of large-scale climate solution approaches, like geoengineering, tended to be male.
22. Aida also argued that women might be more likely to seek the perspective of experts on matters such climate change, saying that "maybe they are better at . . . knowing that they don't know, so they might take the advice of an expert."
23. Interviewees mentioned women's underrepresentation in the scientific community in general, along with specific sectors like the energy or transportation sectors.
24. A few participants apologized for cursing during the interview as well. One said "sorry" three different times during the interview for things like cursing, going off on a tangent, and critiquing a feminist approach that she thought I might be sympathetic to. Another interviewee apologized three times, for saying something in a "long winded way," for saying something that is "quite depressing," and for not providing an answer to my question. Both were native English speakers working in North America.
25. To be fair, I also apologized numerous times in the interviews for things like writing too slowly, not being clear in the questions, or having technical difficulties. In the course of one interview, both the participant and I apologized back-to-back for a miscommunication about when she started working in her current job. There is some scholarly evidence that women do tend to apologize more frequently than men, that they are expected to apologize for offenses, and that they assess perceived offenses as being more severe than men do (Schumann and Ross 2010; Walfisch, Van Dijk, and Kark 2013). Given this, it is perhaps not surprising that there were so many apologies.
26. The IPCC moved to include consideration of gender in its 2013 Fifth Assessment Report (AR5) (Nagel 2016).

27. There have also been publicly discussed instances such as a 2018 IPCC meeting in which the scientists present had to be reminded that this was a meeting of experts and that everyone's expertise must be respected. Some attendees listed a few specific examples of women facing belittling comments or behavior (Yeo 2018).

CHAPTER 6

1. Hurricane Katrina hit the Southeastern part of the United States in 2005. Several states experienced flooding and other negative ramifications of the storm. New Orleans was particularly hard hit and experienced numerous fatalities as well as significant infrastructure and property damage.
2. Interview participants used the women-as-agents discourse 406 times, making it the most widely used discourse of the four explored in the book.
3. For example, Linda Åhäll (2012: 104–105) "explores the workings of power inherent in the way in which female perpetrators of political violence are written as subjects/objects within various discursive practices."
4. Lena Partzsch (2017) contrasts the concept of "power to" with "power over" (coercive power) and "power with" (collective power).
5. The women-as-active-environmental-participants storyline was used sixty-four times.
6. Phillis, a climate scientist in the UK, also discussed speaking at her child's school, but she referred to this as "outreach"—"just teach them the basic physics."
7. According to World Bank data, women actually accounted for 49.6 percent of the world's total population in 2020.
8. This savior storyline often appears alongside the victim storyline discussed in chapter 3. See, for example, Seema Arora-Jonsson (2011) for further discussion.
9. There are also those that point out a lack of data on women's unique climate change experiences (e.g., Rao and Raj 2019), which makes it difficult to even understand the full picture of gendered climate change effects.
10. Five participants expressed skepticism that women would have made fundamentally different decisions about fossil fuel use or other factors relevant to the causes of climate change.
11. An exception is Selma mentioning WEDO by name—"I see all these huge movements like WEDO, etc., where you encounter incredible characters."
12. Sustainable consumption is ubiquitous in "what you can do to help" lists in environmental documentaries, nonprofit factsheets, and other outlets.
13. There were a few exceptions to this trend, including Kathryn who works for a nonprofit in the United States. She recounted an event in her state featuring a woman from the Marshall Islands who is a poet, professor, and climate justice activist and who uses poetry to tell the stories of how climate change is and will continue to harm and impact her community.

14. Joe Curnow and Anjali Helferty (2018) specifically examine the racialized history of the environmental movement in the United States and Canada, and Christopher Carter (2018) focuses on the US context.
15. The study looked at conservation and preservation organizations, government environmental agencies, and environmental grantmaking foundations.
16. There have been several articles and blog posts about the whiteness of environmentalism. See, for example, Dahmen (2017), Swaminathan (2017), and Toomey (2018).

CHAPTER 7

1. The IPCC (2018, 19) warns that "if poorly designed or implemented, adaptation projects in a range of sectors can increase greenhouse gas emissions and water use, increase gender and social inequality, undermine health conditions, and encroach on natural ecosystems."
2. Joe Curnow and Jody R. Chan (2016, 146) studied youth activists in Canada and found that representations of expertise related to three main forms of interaction: "ideas being accepted without discussion, exclusive talk, and affirmation. Through these practices, we see expertise being conflated with dominant masculine forms of participation, making those modes of participation difficult, and sometimes impossible, for women to perform."

References

500 Women Scientists. 2019. Webpage. https://500womenscientists.org.

Abramovitz, Mimi. 1996. *Regulating the Lives of Women: Social Welfare Policy from Colonial Times to the Present.* Boston: South End Press.

Ackerly, Brooke. 2009. "Why a Feminist Theorist Studies Methods." *Politics and Gender* 5 (3): 431–436.

Ackerly, Brooke, Maria Stern, and Jacqui True. 2006. *Feminist Methodologies for International Relations.* New York: Cambridge University Press.

Ackerly, Brooke, and Jacqui True. 2010. *Doing Feminist Research in Political and Social Science.* New York: Palgrave Macmillan.

Adger, W. Neil. 2006. "Vulnerability." *Global Environmental Change* 16 (3): 268–281.

Adger, W. Neil, Hallie Eakin, and Alexandra Winkels. 2009. "Nested and Teleconnected Vulnerabilities to Environmental Change." *Frontiers in Ecology and the Environment* 7 (3): 150–157.

Afifi, Tamer. 2011. "Economic or Environmental Migration? The Push Factors in Niger." *International Migration* 49 (1): e95–e124.

Agarwal, Bina. 2009. "Gender and Forest Conservation: The Impact of Women's Participation in Community Forest Governance." *Ecological Economics* 68 (11): 2785–2799.

Agyeman, Julian, Robert D. Bullard, and Bob Evans, eds. 2003. *Just Sustainabilities: Development in an Unequal World.* Cambridge, MA: MIT Press.

Agyeman, Julian, and Bob Evans. 2004. "'Just Sustainability': The Emerging Discourse of Environmental Justice in Britain?" *Geographical Journal* 170 (2): 155–164.

Åhäll, Linda. 2012. "Motherhood, Myth and Gendered Agency in Political Violence." *International Feminist Journal of Politics* 14 (1): 103–120.

Akinsemolu, Adenike A., and Olukoya A.P. Obafemi. 2020. "The Vulnerability of Women to Climate Change in Coastal Regions of Nigeria: A Case of the Ilaje Community in Ondo State." *Journal of Cleaner Production* 246 (February 10): 119015.

Alaimo, Stacy. 2009. "Insurgent Vulnerability and the Carbon Footprint of Gender." *Women, Gender and Research* 3–4: 22–35.

Alexander, Clarence, Nora Bynum, Elizabeth Johnson, Ursula King, Tero Mustonen, Peter Neofotis, Noel Oettlé, et al. 2011. "Linking Indigenous and Scientific Knowledge of Climate Change." *BioScience* 61 (6): 477–484.

Alscher, Stefan. 2011. "Environmental Degradation and Migration on Hispaniola Island." *International Migration* 49 (1): e95–e124.

Alston, Margaret. 2011. "Gender and Climate Change in Australia." *Journal of Sociology* 47 (1): 53–70.

Alston, Margaret. 2015. *Women and Climate Change in Bangladesh.* New York: Routledge.

American Council on Education. 2017. "American College President Study." https://www.aceacps.org/summary-profile/.

Anastario, Michael P., Ryan Larrance, and Lynn Lawry. 2008. "Using Mental Health Indicators to Identify Postdisaster Gender-Based Violence among Women Displaced by Hurricane Katrina." *Journal of Women's Health* 17(9): 1437–1444.

Andersen, Lykke E., Dorte Verner, and Manfred Wiebelt. 2017. "Gender and Climate Change in Latin America: An Analysis of Vulnerability, Adaptation and Resilience Based on Household Surveys." *Journal of International Development* 29 (7): 857–876.

Anshelm, Jonas, and Martin Hultman. 2014. "A Green Fatwā? Climate Change as a Threat to the Masculinity of Industrial Modernity." *NORMA* 9 (2): 84–96.

Ariyabandu, Madhavi Malalgoda. 2009. "Sex, Gender, and Gender Relations in Disasters." In *Women, Gender, and Disaster: Global Issues and Initiatives*, edited by Elaine Enarson and P. G. Dhar Chakrabarti, 5–18. Los Angeles: Sage.

Arora-Jonsson, Seema. 2009. "Discordant Connections: Discourses on Gender and Grassroots Activism in Two Forest Communities in India and Sweden." *Signs: Journal of Women in Culture and Society* 35 (1): 213–240.

Arora-Jonsson, Seema. 2011. "Virtue and Vulnerability: Discourses on Women, Gender and Climate Change." *Global Environmental Change* 21 (2): 744–751.

Arora-Jonsson, Seema, and Bimbika Basnett Sijapati. 2018. "Disciplining Gender in Environmental Organizations: The Texts and Practices of Gender Mainstreaming." *Gender, Work and Organization* 25 (3): 309–325.

Atteridge, Aaron, and Elise Remling. 2018. "Is Adaptation Reducing Vulnerability or Redistributing It?" *WIRES Climate Change* 9 (1): e500.

Awumbila, Mariama, and Janet Henshall Momsen. 1995. "Gender and the Environment: Women's Time Use as a Measure of Environmental Change." *Global Environmental Change* 5 (4): 337–346.

Axelrod, Mark. 2011. "Climate Change and Global Fisheries: Linking Issues to Protect Ecosystems or to Save Political Interests?" *Global Environmental Politics* 11 (3): 64–84.

Azong, Matilda N., and Clare J. Kelso. 2021. "Gender, Ethnicity and Vulnerability to Climate Change: The Case of Matrilineal and Patrilineal Societies in Bamenda Highlands Region, Cameroon." *Global Environmental Change* 67: 102241.

Bäckstrand, Karin. 2003. "Civic Science for Sustainability: Reframing the Role of Experts, Policy-Makers and Citizens in Environmental Governance." *Global Environmental Politics* 3 (4): 24–41.

Bäckstrand, Karin. 2004. "Scientisation vs. Civic Expertise in Environmental Governance: Eco-Feminist, Eco-Modern and Post-Modern Responses." *Environmental Politics* 13 (4): 695–714.

Bakhtsiyarava, Maryia, Kathryn Grace, and Raphael J. Nawrotzki. 2018. "Climate, Birth Weight, and Agricultural Livelihoods in Kenya and Mali." *American Journal of Public Health* 108 (S2): 144–150.

Barnett, Jon. 2020. "Global Environmental Change II: Political Economies of Vulnerability to Climate Change." *Progress in Human Geography* 44 (6): 1–13.

Barnett, Jon, Simon Lambert, and Ian Fry. 2008. "The Hazards of Indicators: Insights from the Environmental Vulnerability Index." *Annals of the Association of American Geographers* 98 (1): 102–119.

Barnett, Jon, and Saffron O'Neill. 2010. "Maladaptation." *Global Environmental Change* 20 (2): 211–213.

Barr, Stewart, Andrew Gilg, and Gareth Shaw. 2011. "Citizens, Consumers and Sustainability: (Re)Framing Environmental Practice in an Age of Climate Change." *Global Environmental Change* 21 (4): 1224–1233.

Bauhardt, Christine. 2014. "'Solution to the Crisis? The Green New Deal, Degrowth, and the Solidarity Economy: Alternatives to the Capitalist Growth Economy from an Ecofeminist Economics Perspective." *Ecological Economics* 102: 60–68.

Benería, Lourdes. 1995. "Toward a Greater Integration of Gender in Economics." *World Development* 23 (11): 1839–1850.

Benería, Lourdes. 2003. *Gender, Development, and Globalization: Economics as If People Mattered.* London: Routledge.

Benford, Robert D., and David A. Snow. 2000. "Framing Processes and Social Movements: An Overview and Assessment." *Annual Review of Sociology* 26: 611–639.

Bergmann, Zoe, and Ringo Ossewaarde. 2020. "Youth Climate Activists Meet Environmental Governance: Ageist Depictions of the FFF Movement and Greta Thunberg in German Newspaper Coverage." *Journal of Multicultural Discourses* 15 (3): 267–290.

Berry, Helen Louise, Kathryn Bowen, and Tord Kjellstrom. 2010. "Climate Change and Mental Health: A Causal Pathways Framework." *International Journal of Public Health* 55 (2): 123–132.

Bershidsky, Leonid. 2019. "Future Technology Will Solve Climate Change? Don't Believe It." Bloomberg, September 25. https://www.bloomberg.com/opinion/articles/2019-09-26/future-technology-will-solve-climate-change-don-t-believe-it.

Betsill, Michele M. 2014. "Transnational Actors in International Environmental Politics." In *Advances in International Environmental Politics*, edited by Michele M. Betsill, Kathryn Hochstetler, and Dimitris Stevis, 185–210. New York: Palgrave Macmillan.

Bhattarai, Basundhara, Ruth Beilin, and Rebecca Ford. 2015. "Gender, Agrobiodiversity, and Climate Change: A Study of Adaptation Practices in the Nepal Himalayas." *World Development* 70: 122–132.

Biermann, Frank, and Ina Möller. 2019. "Rich Man's Solution? Climate Engineering Discourses and the Marginalization of the Global South." *International Environmental Agreements: Politics, Law, and Economics* 19 (1): 1–17. https://doi.org/https://doi.org/10.1007/s10784-019-09431-0.

Bittman, Michael, Paula England, Nancy Folbre, Liana Sayer, and George Matheson. 2003. "When Does Gender Trump Money? Bargaining and Time in Household Work." *American Journal of Sociology* 109 (1): 186–214.

Blicharska, Malgorzata, Richard Smithers, Magdalena Kuchler, Ganesh Agrawal, José Gutiérrez, Ahmed Hassanali, Saleemul Huq, et al. 2017. "Steps to Overcome the North-South Divide in Research Relevant to Climate-Change Policy and Practice." *Nature Climate Change* 7: 21–27.

Blocker, T. Jean, and Douglas Lee Eckberg. 1989. "Environmental Issues as Women's Issues: General Concerns and Local Hazards." *Social Science Quarterly* 70 (3): 586–593.

Blocker, T. Jean, and Douglas Lee Eckberg. 1997. "Gender and Environmentalism." *Social Science Quarterly* 78 (4): 841–858.

Blum, Elizabeth D. 2008. *Love Canal Revisited: Race, Class, and Gender in Environmental Activism*. Lawrence: University of Kansas Press.

Bohle, Hans-Georg, T. Downing, and Michael J. Watts. 1994. "Climate Change and Social Vulnerability: Toward a Sociology and Geography of Food Insecurity." *Global Environmental Change* 4: 37–48.

Bord, R. J., and R. E. O'Connor. 1997. "The Gender Gap in Environmental Attitudes." *Social Science Quarterly* 78: 830–840.

Bourque, François, and Ashlee Cunsolo Willox. 2014. "Climate Change: The Next Challenge for Public Mental Health?" *International Review of Psychiatry* 26 (4): 415–422.

Boykoff, M., M. Aoyagi, A. Benham, P. Chandler, M. Daly, K. Doi, R. Fernández-Reyes, et al. 2020. "World Newspaper Coverage of Climate Change or Global Warming, 2004–2020." Media and Climate Change Observatory Data Sets. Center for Science and Technology Policy Research, Cooperative Institute for Research in Environmental Sciences, University of Colorado. https://sciencepolicy.colorado.edu/icecaps/research/media_coverage/world/index.html.

Boykoff, Max, and J. Timmons Roberts. 2007. "Media Coverage of Climate Change: Current Trends, Strengths, Weaknesses." UN Human Development Report Office. https://www.researchgate.net/publication/228637999_Media_coverage_of_climate_change_Current_trends_strengths_weaknesses.

Braun, Yvonne A. 2011. "Left High and Dry: An Intersectional Analysis of Gender, Dams and Development in Lesotho." *International Feminist Journal of Politics* 13 (2): 141–162.

Bretherton, Charlotte. 2003. "Movements, Networks, Hierarchies: A Gender Perspective on Global Environmental Governance." *Global Environmental Politics* 3 (2): 103–119.

Brklacich, Mike, May Chazan, and Hans-Georg Bohle. 2010. "Human Security, Vulnerability, and Global Environmental Change." In *Global Environmental Change and Human Security*, edited by Richard A. Matthew, Jon Barnett, Bryan McDonald, and Karen L. O'Brien, 35–52. Cambridge, MA: MIT Press.

Brough, Aaron R., James E. B. Wilkie, Jingjing Ma, Mathew S. Isaac, and David Gal. 2016. "Is Eco-Friendly Unmanly? The Green-Feminine Stereotype and Its Effect on Sustainable Consumption." *Journal of Consumer Research* 43 (4): 567–582.

Brown, Katrina, and Elizabeth Westaway. 2011. "Agency, Capacity, and Resilience to Environmental Change: Lessons from Human Development, Well-Being, and Disasters." *Annual Review of Environment and Resources* 36 (1): 321–342.

Buck, Holly Jean, Andrea R. Gammon, and Christopher J. Preston. 2014. "Gender and Geoengineering." *Hypatia* 29 (3): 651–669.

Buckingham, Susan. 2020. *Gender and Environment.* New York: Routledge.

Bullard, Robert D. 2005. *The Quest for Environmental Justice: Human Rights and the Politics of Pollution*. San Francisco, CA: Sierra Club Books.

Bullard, Robert D., and Damu Smith. 2005. "Women Warriors of Color on the Front Line." In *The Quest for Environmental Justice*, edited by Robert D. Bullard, 62–84. San Francisco, CA: Sierra Club Books.

Bullard, Robert D., and Beverly Wright. 2012. *The Wrong Complexion for Protection: How the Government Response to Disaster Endangers African American Communities.* New York: New York University Press.

Burton, Nylah. 2019. "Meet the Young Activists of Color Who Are Leading the Charge against Climate Disaster." *Vox*, October 11. https://www.vox.com/identities/2019/10/11/20904791/young-climate-activists-of-color.

Butler, Cameron. 2017. "A Fruitless Endeavour: Confronting the Heteronormativity of Environmentalism." In *Routledge Handbook of Gender and Environment*, edited by Sherilyn MacGregor, 270–286. New York: Routledge.

Butler, Judith. 2006. *Precarious Life: The Powers of Mourning and Violence*. London: Verso.

Butler, Judith. 2020. *The Force of Nonviolence: The Ethical in the Political.* New York: Verso.

Calma, Justine, and Paola Rosa-Aquino. 2019. "4 Black Women Leaders on Climate, Justice, and the Green 'Promised Land.'" *Grist* (blog), February 27. https://grist.org/article/4-black-women-leaders-on-climate-justice-and-the-green-promised-land.

Cannon, Terry. 2002. "Gender and Climate Hazards in Bangladesh." *Gender and Development* 10 (2): 45–50.

Carreon, Michelle E., and Valentine M. Moghadam. 2015. "'Resistance Is Fertile': Revisiting Maternalist Frames across Cases of Women's Mobilization." *Women's Studies International Forum* 51 (Jul–Aug): 19–30.

Carrigan, Anthony. 2010. "Postcolonial Disaster, Pacific Nuclearization, and Disabling." *Journal of Literary & Cultural Disability Studies* 4 (3): 255–272.

Carter, Christopher. 2018. "Blood in the Soil: The Racial, Racist, and Religious Dimensions of Environmentalism." In *The Bloomsbury Handbook of Religion and Nature: The Elements*, edited by Laura Hobgood and Whitney Bauman, 45–62. New York: Bloomsbury Academic.

Castillo, Rosalva Aída Hernández. 2008. "Gendered Violence and Neocolonialism: Indigenous Women Confronting Counterinsurgency Violence." *Latin American Perspectives* 35 (1): 151–54.

Causevic, Az, Kavita Philip, Maari Zwick-Maitreyi, Persephone Hooper Lewis, Siko Bouterse, and Anasuya Sengupta. 2020. "Centering Knowledge from the Margins: Our Embodied Practices of Epistemic Resistance and Revolution." *International Feminist Journal of Politics* 22 (1): 6–25.

Cech, Erin A., and Mary Blair-Loy. 2019. "The Changing Career Trajectories of New Parents in STEM." *Proceedings of the National Academy of Sciences of the United States of America*, 116 (10): 4182–4187. https://www.pnas.org/content/early/2019/02/12/1810862116.

Center for International Environmental Law. 2019. "Fuel to the Fire: How Geoengineering Threatens to Entrench Fossil Fuels and Accelerate the Climate Crisis." Washington, DC. https://www.ciel.org/wp-content/uploads/2019/02/CIEL_FUEL-TO-THE-FIRE_How-Geoengineering-Threatens-to-Entrench-Fossil-Fuels-and-Accelerate-the-Climate-Crisis_February-2019.pdf.

Center for Naval Analyses. 2007. "National Security and the Threat of Climate Change." https://www.cna.org/sites/default/files/National%20Security%20and%20the%20Threat%20of%20Climate%20Change%20-%20Print.pdf.

Challinor, Andrew, Tim Wheeler, Chris Garforth, Peter Craufurd, and Amir Kassam. 2007. "Assessing the Vulnerability of Food Crop Systems in Africa to Climate Change." *Climatic Change* 83 (3): 381–399.

Cimpian, Andrei, and Sarah-Jane Leslie. 2017. "The Brilliance Paradox: What Really Keeps Women and Minorities from Excelling in Academia." *Scientific American*, September 14. https://www.scientificamerican.com/article/the-brilliance-paradox-what-really-keeps-women-and-minorities-from-excelling-in-academia.

Coffey, Amanda, and Paul Atkinson. 1996. *Making Sense of Qualitative Data: Complementary Research Strategies*. London: Sage.

Coffey, Clare, Patricia Espinoza Revollo, Rowan Harvey, Max Lawson, Anam Parvez Butt, Kim Piaget, Diana Sarosi, and Julie Thekkudan. 2020. "Time to Care: Unpaid and Underpaid Care Work and the Global Inequality Crisis." Oxfam. https://ousweb-prodv2-shared-media.s3.amazonaws.com/media/documents/FINAL_bp-time-to-care-inequality-200120-en.pdf.

Cohen, Marjorie Griffin, ed. 2017. *Climate Change and Gender in Rich Countries: Work, Public Policy and Action*. New York: Routledge.

Conliffe, Alexandra. 2011. "Combating Ineffectiveness: Climate Change Bandwagonning and the UN Convention to Combat Desertification." *Global Environmental Politics* 11 (3): 44–63.

Cook, John, Naomi Oreskes, Peter T. Doran, William R. L. Anderegg, Bart Verheggen, Ed W. Maibach, J. Stuart Carlton, et al. 2016. "Consensus on Consensus: A Synthesis of Consensus Estimates on Human-Caused Global Warming." *Environmental Research Letters* 11 (4): 1–7.

Costa Pinto, Diego, Márcia M. Herter, Patricia Rossi, and Adilson Borges. 2014. "Going Green for Self or for Others? Gender and Identity Salience Effects on Sustainable Consumption." *International Journal of Consumer Studies* 38 (5): 540–549.

Cox, Robert W. 1981. "Social Forces, States and World Orders: Beyond International Relations Theory." *Millennium: Journal of International Studies* 10 (2): 128–137.

Crenshaw, Kimberlé. 1989. "Demarginalizing the Intersection of Race and Sex: A Black Feminist Critique of Antidiscrimination Doctrine, Feminist Theory and Antiracist Politics." *University of Chicago Legal Forum* 1989 (8): 139–167.

Crenshaw, Kimberlé. 1994. "Mapping the Margins: Intersectionality, Identity Politics, and Violence against Women of Colour." In *The Public Nature of Private Violence*, edited by M. A. Fineman and R. Mykitiul, 93–120. New York: Routledge.

Cripps, Elizabeth. 2015. "Population and Environment: The Impossible, the Impermissible, and the Imperative." In *The Oxford Handbook of Environmental Ethics*, edited by Stephen Gardiner and Allen Thompson, 1–14. Oxford: Oxford University Press.

Cunsolo, Ashlee, and Neville R. Ellis. 2018. "Ecological Grief as a Mental Health Response to Climate Change-Related Loss." *Nature Climate Change* 8: 275–281.

Cuomo, Chris J. 2011. "Climate Change, Vulnerability, and Responsibility." *Hypatia* 26 (4): 690–714.

Curnow, Joe, and Jody R. Chan. 2016. "Becoming an 'Expert': Gendered Positioning, Praise, and Participation in an Activist Community." In *Transforming Learning, Empowering Learners: The International Conference of the Learning Sciences 2016*, edited by C. K. Looi, J. L. Polman, U. Cress, and P. Reimann, 146–153. Singapore: International Society of the Learning Sciences.

Curnow, Joe, and Anjali Helferty. 2018. "Contradictions of Solidarity: Whiteness, Settler Coloniality, and the Mainstream Environmental Movement." *Environment and Society* 9 (1): 145–163.

Cutter, Susan L. 2006. *Hazards, Vulnerability and Environmental Justice*. New York: Routledge.

Dahmen, Nicole Smith. 2017. "The Overwhelming Whiteness of US Environmentalism Is Hobbling the Fight against Climate Change." *Quartz* (blog), January 4. https://qz.com/877447/the-overwhelming-whiteness-of-the-us-environmentalist-movement-is-hobbling-the-fight-against-climate-change/.

Daly, Mary. 2000. *The Gender Division of Welfare: The Impact of the British and German Welfare States*. Cambridge: Cambridge University Press.

D'Amato, Gennaro, Stephen T. Holgate, Ruby Pawankar, Dennis K. Ledford, Lorenzo Cecchi, Mona Al-Ahmad, Fatma Al-Enezi, et al. 2015. "Meteorological Conditions, Climate Change, New

Emerging Factors, and Asthma and Related Allergic Disorders. A Statement of the World Allergy Organization." *World Allergy Organization Journal* 8 (25): 1–52.

Dankelman, Irene. 2002. "Climate Change: Learning from Gender Analysis and Women's Experiences of Organising for Sustainable Development." In *Gender, Development, and Climate Change*, edited by Rachel Masika, 21–29. Oxford: Oxfam GB.

Dankelman, Irene. 2010. *Gender and Climate Change: An Introduction*. Sterling, VA: Earthscan.

Dauvergne, Peter. 2005. "Dying of Consumption: Accidents or Sacrifices of Global Morality?" *Global Environmental Politics* 5 (3): 35–47.

Dauvergne, Peter. 2008. *The Shadows of Consumption: Consequences for the Global Environment*. Cambridge, MA: MIT Press.

Dauvergne, Peter. 2010. "The Problem of Consumption." *Global Environmental Politics* 10 (2): 1–10.

Dauvergne, Peter, and Jennifer Clapp. 2016. "Researching Global Environmental Politics in the 21st Century." *Global Environmental Politics* 16 (1): 1–12.

David, Emmanuel, and Elaine Enarson, eds. 2012. *The Women of Katrina: How Gender, Race, and Class Matter in an American Disaster.* Nashville, TN: Vanderbilt University Press.

Davidson, Debra J., and William R. Freudenburg. 1996. "Gender and Environmental Risk Concerns." *Environment and Behavior* 28 (3): 302–339.

De Groot, Judith I. M., and Linda Steg. 2009. "Morality and Prosocial Behavior: The Role of Awareness, Responsibility, and Norms in the Norm Activation Model." *Journal of Social Psychology* 149 (4): 425–449.

Demetriades, Justina, and Emily Esplen. 2010. "The Gender Dimensions of Poverty and Climate Change Adaptation." In *Social Dimensions of Climate Change: Equity and Vulnerability in a Warming World*, edited by Robin Mearns and Andrew Norton. Washington, DC: The World Bank, 133–144.

Denton, Fatma. 2002. "Climate Change Vulnerability, Impacts, and Adaptation: Why Does Gender Matter?" *Gender and Development* 10 (2): 10–20.

Deschênes, Olivier, Michael Greenstone, and Jonathan Guryan. 2009. "Climate Change and Birth Weight." *American Economic Review* 99 (2): 211–217.

DeSombre, Elizabeth R. 2018. "Individual Behavior and Global Environmental Problems." *Global Environmental Politics* 18 (1): 5–12.

Detraz, Nicole. 2021. "Gender and Comparative Environmental Politics: Examining Population Debates through Gender Lenses." In *Oxford Handbook on Comparative Environmental Politics*, edited by Jeannie Sowers, Stacy Vandeveer, and Erika Weinthal. Oxford: Oxford University Press. https://www.oxfordhandbooks.com/view/10.1093/oxfordhb/9780197515037.001.0001/oxfordhb-9780197515037.

Detraz, Nicole. 2017a. "Gender and Environmental (In)Security: From Climate Conflict to Ecosystem Instability." In *Routledge International Handbook on Gender and Environment*, edited by Sherilyn MacGregor, 202–215. New York: Routledge.

Detraz, Nicole. 2017b. *Gender and the Environment.* Malden, MA: Polity Press.

Detraz, Nicole, and Dursun Peksen. 2017. "In the Aftermath of Earth, Wind and Fire: Natural Disasters and Respect for Women's Rights." *Human Rights Review* 18 (2): 151–170.

Detraz, Nicole, and Dursun Peksen. 2018. "'Women Friendly' Spending? Welfare Spending and Women's Participation in the Economy and Politics." *Politics and Gender* 14 (2): 137–161.

Detraz, Nicole, and Leah Windsor. 2014. "Evaluating Climate Migration: Population Movement, Insecurity and Gender." *International Feminist Journal of Politics* 16 (1): 127–146.

Dey, Anamika, Gurdeep Singh, and Anil K. Gupta. 2018. "Women and Climate Stress: Role Reversal from Beneficiaries to Expert Participants." *World Development* 103: 336–359.

Dhaliwal, Suzanne. 2015. "Why Are Britain's Green Movements an All-White Affair?" *The Guardian*, September 28. https://www.theguardian.com/environment/2015/sep/28/why-are-britains-green-movements-an-all-white-affair.

Di Chiro, Giovanna. 2008. "Living Environmentalisms: Coalition Politics, Social Reproduction and Environmental Justice." *Environmental Politics* 17 (2): 276–298.

Di Chiro, Giovanna. 2019. "Care Not Growth: Imagining a Subsistence Economy for All." *British Journal of Politics and International Relations* 21 (2): 303–311.

Dillon, Lindsey, and Julie Sze. 2016. "Police Power and Particulate Matters: Environmental Justice and the Spatialities of In/Securities in U.S. Cities." *English Language Notes* 54 (2): 13–23.

Djoudi, Houria, Bruno Locatelli, Chloe Vaast, Kiran Asher, Maria Brockhaus, and Bimbika Basnett Sijapati. 2016. "Beyond Dichotomies: Gender and Intersecting inequalities in Climate Change Studies." *Ambio* 45: S248–S262.

Doherty, Brian, and Timothy Doyle. 2006. "Beyond Borders: Transnational Social Movements and Modern Environmentalisms." *Environmental Politics* 15 (5): 697–712.

Donovan, Josephine, and Carol J. Adams, eds. 1996. *Beyond Animal Rights: A Feminist Caring Ethic for the Treatment of Animals.* New York: Continuum.

Douglas, Susan, and Meredith Michaels. 2005. *The Mommy Myth: The Idealization of Motherhood and How It Has Undermined All Women.* New York: Free Press.

Dove, Michael R. 2006. "Indigenous People and Environmental Politics." *Annual Review of Anthropology* 35: 191–208.

Duncanson, Claire. 2009. "Narratives of Military Masculinity in Peacekeeping Operations." *International Feminist Journal of Politics* 11 (1): 63–80.

Dymén, Christian, and Richard Langlais. 2017. "Integrating Gender and Planning towards Climate Change Response: Theorising from the Swedish Case." In *Understanding Climate Change through Gender Relations*, edited by Susan Buckingham and Virginie Le Masson, 239–256. New York: Routledge.

Eakin, Hallie, and Maggie L. Walser. 2007. "Human Vulnerability to Global Environmental Change." Edited by C. J. Cleveland. Environmental Information Coalition, National Council of Science and the Environment.

Eastin, Joshua. 2018. "Climate Change and Gender Equality in Developing States." *World Development* 107 (July): 289–305.

Eckersley, Robin. 2020. "Ecological Democracy and the Rise and Decline of Liberal Democracy: Looking Back, Looking Forward." *Environmental Politics* 29 (2): 214–234.

Eden, Sally, Andrew Donaldson, and Gordon Walker. 2006. "Green Groups and Grey Areas: Scientific Boundary-Work, Nongovernmental Organizations and Environmental Knowledge." *Environment and Planning A* 38: 1061–1076.

Elborgh-Woytek, Katrin, Monique Newiak, Kalpana Kochhar, Stefania Fabrizio, Kangni Kpodar, Philippe Wingender, Benedict Clements, and Gerd Schwartz. 2013. "Women, Work, and the Economy: Macroeconomic Gains from Gender Equity." International Monetary Fund. http://www.imf.org/external/pubs/ft/sdn/2013/sdn1310.pdf.

El-Bushra, Judy. 2007. "Feminism, Gender, and Women's Peace Activism." *Development and Change* 38 (1): 131–147.

Elias, Juanita, and Adrienne Roberts. 2016. "Feminist Global Political Economies of the Everyday: From Bananas to Bingo." *Globalizations* 13 (6): 787–800.

Enarson, Elaine, and Betty Hearn Morrow. 1998. *The Gendered Terrain of Disaster: Through Women's Eyes*. Westport, CT: Praeger.

Enarson, Elaine, and Bob Pease, eds. 2016. *Men, Masculinities and Disaster*. New York: Routledge.

Enloe, Cynthia. 1990. *Bananas, Beaches and Bases: Making Feminist Sense of International Politics*. Berkeley: University of California Press.

Enloe, Cynthia. 2004. *The Curious Feminist: Searching for Women in a New Age of Empire*. Berkeley: University of California Press.

Enloe, Cynthia. 2016. "Afterword: Being Reflexively Feminist Shouldn't Be Easy." In *Researching War: Feminist Methods, Ethics and Politics*, edited by Annick T. R. Wibben, 258–259. New York: Routledge.

Epstein, Charlotte. 2008. *The Power of Words in International Relations: Birth of an Anti-Whaling Discourse*. Cambridge, MA: MIT Press.

European Commission. 2017. "Special Eurobarometer 459: Climate Change." https://ec.europa.eu/clima/sites/clima/files/support/docs/report_2017_en.pdf.

Evans, Gary W., Siegmar Otto, and Florian G. Kaiser. 2018. "Childhood Origins of Young Adult Behavior." *Psychological Science* 29 (5): 679–687.

Evertsen, Kathinka Fossum, and Kees van der Geest. 2020. "Gender, Environment and Migration in Bangladesh." *Climate and Development* 12 (1): 12–22.

Federici, Silvia. 2009. "The Devaluation of Women's Labour." In *Eco-Sufficiency and Global Justice: Women Write Political Ecology*, edited by Ariel Salleh, 43–65. New York: Pluto Press.

Feminist Green New Deal. 2019. "Principles." http://feministgreennewdeal.com/principles.

Ferguson, Lucy. 2015. "'This Is Our Gender Person': The Messy Business of Working as a Gender Expert in International Development." *International Feminist Journal of Politics* 17 (3): 380–397.

First, Jennifer M., Nathan L. First, and J. Brian Houston. 2017. "Intimate Partner Violence and Disasters: A Framework for Empowering Women Experiencing Violence in Disaster Settings." *Affilia* 32 (3): 390–403.

Fischer, Anke, Vera Peters, Mirjam Neebe, Jan Vávra, Antoinette Kriel, Miloslav Lapka, and Boldizsár Megyesi. 2012. "Climate Change? No, Wise Resource Use Is the Issue: Social Representations of Energy, Climate Change and the Future." *Environmental Policy Governance* 22 (3): 161–176.

Fisher, Sarah. 2010. "Violence against Women and Natural Disasters: Findings from Post-Tsunami Sri Lanka." *Violence Against Women* 16 (8): 902–918.

Ford, Lucy H. 2003. "Challenging Global Environmental Governance: Social Movement Agency and Global Civil Society." *Global Environmental Politics* 3 (2): 120–134.

Formetta, Giuseppe, and Luc Feyen. 2019. "Empirical Evidence of Declining Global Vulnerability to Climate-Related Hazards." *Global Environmental Change* 57: 101920.

Foster, Emma A. 2017. "Gender, Environmental Governmentality, and the Discourses of Sustainable Development." In *Routledge Handbook of Gender and Environment*, edited by Sherilyn MacGregor, 216–228. New York: Routledge.

Fothergill, Alice, and Lori Peek. 2015. *Children of Katrina*. Austin: University of Texas Press.

Fox, Mary Frank, and Glenn Firebaugh. 1992. "Confidence in Science." *Social Science Quarterly* 73 (1): 101–113.

Fox Keller, Evelyn. 1995. *Reflections on Gender and Science*. New Haven, CT: Yale University Press.

Fraser, Nancy. 1987. "Women, Welfare and the Politics of Need Interpretation." *Hypatia* 2 (1): 103–121.

Friedman, Marilyn. 1987. "Beyond Caring: The De-Moralization of Gender." *Canadian Journal of Philosophy* 17 (1): 87–110.

Fuchs, Britta, Doris Damyanovic, Karin Weber, and Florian Reinwald. 2017. "A Gender-Sensitive Analysis of Spatial Planning Instruments Related to the Management of Natural Hazards in Austria." In *Understanding Climate Change through Gender Relations*, edited by Susan Buckingham and Virginie Le Masson, 257–275. New York: Routledge.

Fuchs, Doris A., and Sylvia Lorek. 2005. "Sustainable Consumption Governance: A History of Promises and Failures." *Journal of Consumer Policy* 28 (3): 261–288.

Furusa, Zanele, and Munashe Furusa. 2014. "Women's Coping and Adaptation Capacities in Pastoralist Communities in Africa: Dealing with Climate Variability and Change." *Agenda* 28 (3): 65–72.

Füssel, Hans-Martin, and Richard J. T. Klein. 2006. "Climate Change Vulnerability Assessments: An Evolution of Conceptual Thinking." *Climatic Change* 75 (3): 301–329.

Gaard, Greta. 1997. "Toward a Queer Ecofeminism." *Hypatia* 12 (1): 114–137.

Gaard, Greta. 2011. "Ecofeminism Revisited: Rejecting Essentialism and Re-Placing Species in a Material Feminist Environmentalism." *Feminist Formations* 23 (2): 26–53.

Gaillard, J. C. 2010. "Vulnerability, Capacity and Resilience: Perspectives for Climate and Development Policy." *Journal of International Development* 22 (2): 218–232.

Gay-Antaki, Miriam, and Diana Liverman. 2018. "Women Scientists and the IPCC." *Proceedings of the National Academy of Sciences* 115 (9): 2060–2065. https://doi.org/10.1073/pnas.1710271115.

Gelin, Martin. 2019. "The Misogyny of Climate Deniers." *New Republic*, August 28. https://newrepublic.com/article/154879/misogyny-climate-deniers.

Gentry, Caron E. 2009. "Twisted Maternalism: From Peace to Violence." *International Feminist Journal of Politics* 11 (2): 235–252.

Gentry, Caron. 2016a. "Chechen Political Violence as Desperation: What Feminist Discourse Analysis Reveals." In *Researching War: Feminist Methods, Ethics and Politics*, edited by Annick T. R. Wibben, 19–37. New York: Routledge.

Gentry, Caron E. 2016b. "Feminist Christian Realism: Vulnerability, Obligation and Power Politics." *International Feminist Journal of Politics* 18 (3): 449–467.

Gentry, Caron E., and Laura Sjoberg. 2015. *Beyond Mothers, Monsters, Whores: Thinking about Women's Violence in Global Politics*. New York: Zed Books.

Ghodsee, Kristen. 2007. "Potions, Lotions and Lipstick: The Gendered Consumption of Cosmetics and Perfumery in Socialist and Post-Socialist Urban Bulgaria." *Women's Studies International Forum* 30 (1): 26–39.

Glazebrook, Trish. 2011. "Women and Climate Change: A Case-Study from Northeast Ghana." *Hypatia* 26 (4): 762–782.

Glazebrook, Trish, and Emmanuela Opoku. 2018. "Defending the Defenders: Environmental Protectors, Climate Change and Human Rights." *Ethics and the Environment* 23 (2): 83–109.

Global Witness. 2019. "Enemies of the State?" https://www.globalwitness.org/en/campaigns/environmental-activists/enemies-state.

Gonda, Noémi. 2017. "Revealing the Patriarchal Sides of Climate Change Adaptation through Intersectionality: A Case Study from Nicaragua." In *Understanding Climate Change through Gender Relations*, edited by Susan Buckingham and Virginie Le Masson, 173–189. New York: Routledge.

Gottlieb, Robert. 2005. *Forcing the Spring: The Transformation of the American Environmental Movement*. Washington, DC: Island Press.

Green, D., and G. Raygorodetsky. 2010. "Indigenous Knowledge of a Changing Climate." *Climatic Change* 100 (2): 239–242.

Green, Jessica F., and Thomas N. Hale. 2017. "Reversing the Marginalization of Global Environmental Politics in International Relations: An Opportunity for the Discipline." *PS: Political Science and Politics* 50 (2): 473–479.

Grønhøj, Alice, and John Thøgersen. 2012. "Action Speaks Louder than Words: The Effect of Personal Attitudes and Family Norms on Adolescents' Pro-Environmental Behaviour." *Journal of Economic Psychology* 33 (1): 292–302.

Guy, Sophie, Yoshihisa Kashima, Iain Walker, and Saffron O'Neill. 2014. "Investigating the Effects of Knowledge and Ideology on Climate Change Beliefs." *European Journal of Social Psychology* 44 (5): 421–429.

Haas, Peter M., and Ronald B. Mitchell. 2013. "Advanced Scholarship: Interdisciplinary Research at the Science-Policy Interface." In *Routledge Handbook of Global Environmental Politics*, edited by Paul G. Harris, 72–82. New York: Routledge.

Hajer, Maarten. 1995. *The Politics of Environmental Discourse: Ecological Modernization and the Policy Process*. London: Oxford University Press.

Hamilton, Lawrence C. 1985a. "Concern about Toxic Wastes: Three Demographic Predictors." *Sociological Perspectives* 28: 463–486.

Hamilton, Lawrence C. 1985b. "Who Cares about Water Pollution: Opinions in a Small-Town Crisis." *Sociological Inquiry* 55 (2): 170–181.

Harding, Sandra. 1991. *Whose Science? Whose Knowledge? Thinking from Women's Lives*. Ithaca, NY: Cornell University Press.

Harding, Sandra. 1993. *The Science Question in Feminism*. Ithaca, NY: Cornell University Press.

Hardoy, Jorgelina, and Gustavo Pandiella. 2009. "Urban Poverty and Vulnerability to Climate Change in Latin America." *Environment and Urbanization* 21 (1): 203–224.

Hartmann, Betsy. 2010. "Rethinking the Role of Population in Human Security." In *Global Environmental Change and Human Security*, edited by Richard A. Matthew, Jon Barnett, Bryan McDonald, and Karen L. O'Brien, 193–214. Cambridge, MA: MIT Press.

Hartmann, Betsy. 2014. "Converging on Disaster: Climate Security and the Malthusian Anticipatory Regime for Africa." *Geopolitics* 19 (4): 757–783.

Harvey, Fiona. 2018. "Mary Robinson Launches New Feminist Fight against Climate Change." *The Guardian*, July 24. https://www.theguardian.com/environment/2018/jul/24/mary-robinson-launches-new-feminist-fight-against-climate-change.

Haynes, Alex. 2017. "Everyday Life in Rural Bangladesh: Understanding Gender Relations in the Context of Climate Change." In *Understanding Climate Change through Gender Relations*, edited by Susan Buckingham and Virginie Le Masson, 141–156. New York: Routledge.

Hemmati, Minu, and Ulrike Röhr. 2009. "Engendering the Climate-Change Negotiations: Experiences, Challenges, and Steps Forward." *Gender and Development* 17 (1): 19–32.

Hertel, Thomas W., Marshall B. Burke, and David B. Lobell. 2010. "The Poverty Implications of Climate-Induced Crop Yield Changes by 2030." *Global Environmental Change* 20 (4): 577–585.

Hochstetler, Kathryn, and Melinda Laituri. 2014. "Methods in International Environmental Politics." In *Advances in International Environmental Politics*, edited by Michele Betsill, Kathryn Hochstetler, and Dimitris Stevis, 78–106. New York: Palgrave Macmillan.

Hopwood, Bill, Mary Mellor, and Geoff O'Brien. 2005. "Sustainable Development: Mapping Different Approaches." *Sustainable Development* 13: 38–52.

Howell, Junia, and James R. Elliott. 2019. "Damages Done: The Longitudinal Impacts of Natural Hazards on Wealth Inequality in the United States." *Social Problems* 66 (3): 448–467. https://doi.org/10.1093/socpro/spy016.

Hultman, Martin. 2017. "Exploring Industrial, Ecomodern, and Ecological Masculinities." In *Routledge Handbook of Gender and Environment*, edited by Sherilyn MacGregor, 239–252. New York: Routledge.

Hultman, Martin, and Jonas Anshelm. 2017. "Masculinities of Global Climate Change: Exploring Ecomodern, Industrial, and Ecological Masculinity." In *Climate Change and Gender in Rich Countries: Work, Public Policy and Action*, edited by Marjorie Griffin Cohen, 19–34. New York: Routledge.

Humbert, Anne Laure, Elisabeth Kelan, and Marieke van den Brink. 2019. "The Perils of Gender Beliefs for Men Leaders as Change Agents for Gender Equality." *European Management Review* 16 (4): 1143–1157.

Hunt, Elle. 2020. "The Eco Gender Gap: Why Is Saving the Planet Seen as Women's Work?" *The Guardian*, February 6. https://www.theguardian.com/environment/2020/feb/06/eco-gender-gap-why-saving-planet-seen-womens-work.

Hunter, L. M., A. Hatch, and A. Johnson. 2004. "Cross-National Gender Variation in Environmental Behaviors." *Social Science Quarterly* 85 (3): 677–694.

Huynh, Phuong T. A., and Bernadette P. Resurrección. 2014. "Women's Differentiated Vulnerability and Adaptations to Climate-Related Agricultural Water Scarcity in Rural Central Vietnam." *Climate and Development* 6 (3): 226–237.

Ignatow, Gabriel. 2005. "From Science to Multiculturalism: Postmodern Trends in Environmental Organizations." *Global Environmental Politics* 5 (2): 88–113.

Intergovernmental Panel on Climate Change. 2014a. *Climate Change 2014: Impacts, Adaptation, and Vulnerability. Part A: Global and Sectoral Aspects. Contribution of Working Group II to the Fifth Assessment Report of the Intergovernmental Panel on Climate Change.* New York: Cambridge University Press. http://ipcc.ch/pdf/assessment-report/ar5/wg2/WGIIAR5-PartA_FINAL.pdf.

Intergovernmental Panel on Climate Change. 2014b. "Climate Change 2014: Synthesis Report. Contribution of Working Groups I, II and III to the Fifth Assessment Report of the Intergovernmental Panel on Climate Change." IPCC. https://www.ipcc.ch/pdf/assessment-report/ar5/syr/SYR_AR5_FINAL_full_wcover.pdf.

Intergovernmental Panel on Climate Change. 2014c. "Summary for Policymakers." https://www.ipcc.ch/site/assets/uploads/2018/02/ar5_wgII_spm_en.pdf.

Intergovernmental Panel on Climate Change. 2018. "Global Warming of 1.5°C, an IPCC Special Report on the Impacts of Global Warming of 1.5°C above Pre-Industrial Levels and Related Global Greenhouse Gas Emission Pathways, in the Context of Strengthening the Global Response to the Threat of Climate Change, Sustainable Development, and Efforts to Eradicate Poverty." https://www.ipcc.ch/sr15/.

IPCC. *See* Intergovernmental Panel on Climate Change.

Israel, Andrei L., and Carolyn Sachs. 2012. "A Climate for Feminist Intervention: Feminist Science Studies and Climate Change." In *Research, Action and Policy: Addressing the Gendered Impacts of Climate Change*, edited by Margaret Alston and Kerri Whittenbury, 33–51. New York: Springer.

Ivanova, Maria. 2015. "COP21: Why More Women Need Seats at the Table." CNN, 2015. http://www.cnn.com/2015/11/23/opinions/cop21-women-need-participate/.

Jacobs, Janis E., and Sandra D. Simpkins. 2006. *Leaks in the Pipeline to Math, Science, and Technology Careers.* San Francisco: Jossey-Bass.

Jaggar, Alison M. ed. 2014. *Gender and Global Justice*. Malden, MA: Polity Press.

Javeline, Debra. 2014. "The Most Important Topic Political Scientists Are Not Studying: Adapting to Climate Change." *Perspectives on Politics* 12 (2): 420–434.

Jinnah, Sikina. 2011. "Climate Change Bandwagonning: The Impacts of Strategic Linkages on Regime Design, Maintenance, and Death." *Global Environmental Politics* 11 (3): 1–9.

Jinnah, Sikina. 2018. "Why Govern Climate Engineering? A Preliminary Framework for Demand-Based Governance." *International Studies Review* 20: 272–282.

Jinnah, Sikina, and Simon Nicholson. 2019. "Governing Solar Radiation Management: How, Why, and for Whom?" *Environmental Politics* 28 (3): 385–396.

Johnson, Paula A., Sheila E. Widnall, and Frazier F. Benya, eds. 2018. "Sexual Harassment of Women: Climate, Culture, and Consequences in Academic Sciences, Engineering, and Medicine." Washington, DC: National Academies of Sciences, Engineering, and Medicine. https://www.nap.edu/read/24994/chapter/1.

Johnsson-Latham, Gerd. 2006. "Initial Study of Lifestyles, Consumption Patterns, Sustainable Development and Gender: Do Women Leave a Smaller Ecological Footprint than Men?" Swedish Ministry of Sustainable Development. http://www.government.se/content/1/c6/06/72/73/1bd9aa9c.pdf.

Jones, Karen R. 2019. "From William Wordsworth to Extinction Rebellion: A History of Britain's Green Activists." *History Extra* (blog), October 22. https://www.historyextra.com/period/modern/william-wordsworth-extinction-rebellion-history-britains-green-activists-environment-campaigning-activism-environmentalism.

Jylhä, Kirsti M., Clara Cantal, Nazar Akrami, and Taciano Milfont. 2016. "Denial of Anthropogenic Climate Change: Social Dominance Orientation Explains the Conservative Male Effect in Brazil and Sweden." *Personality and Individual Differences* 98: 184–187.

Kaijser, Anna, and Annica Kronsell. 2014. "Climate Change through the Lens of Intersectionality." *Environmental Politics* 23 (3): 417–433.

Kelan, Elisabeth. 2009. *Performing Gender at Work.* London: Palgrave Macmillan.

Kelly, P. M., and W. Neill Adger. 2000. "Theory and Practice in Assessing Vulnerability to Climate Change and Facilitating Adaptation." *Climatic Change* 47 (4): 325–352.

Kennedy, Emily Huddart, and Julie Kmec. 2018. "Reinterpreting the Gender Gap in Household Pro-Environmental Behaviour." *Environmental Sociology* 4 (3): 299–310.

Kenny, Meryl, and Fiona Mackay. 2018. "Feminist and Gendered Approaches." In *Theory and Methods in Political Science,* edited by Vivien Lowndes, David Marsh, and Gerry Stoker, 4th edition, 92–108. New York: Palgrave.

Kent, Jennifer. 2009. "Individualized Responsibility and Climate Change: 'If Climate Protection Becomes Everyone's Responsibility, Does It End up Being No-One's?'" *Cosmopolitan Civil Societies* 1 (3): 132–149.

Keohane, Robert O. 1998. "Beyond Dichotomy: Conversations between International Relations and Feminist Theory." *International Studies Quarterly* 42 (1): 193–197.

Keohane, Robert O. 2015. "The Global Politics of Climate Change: Challenge for Political Science." *PS: Political Science and Politics* 48 (1): 19–26.

Kersey, Alyssa J., Emily J. Braham, Kelsey D. Csumitta, Melissa E. Libertus, and Jessica F. Cantlon. 2018. "No Intrinsic Gender Differences in Children's Earliest Numerical Abilities." *NPJ Science of Learning* 3 (12): 1–10. https://doi.org/10.1038/s41539-018-0028-7.

Kersey, Alyssa J., Kelsey D. Csumitta, and Jessica F. Cantlon. 2019. "Gender Similarities in the Brain during Mathematics Development." *NPJ Science of Learning* 4 (19): 1–7. https://doi.org/10.1038/s41539-019-0057-x.

Khan, Aneire Ehmar, Andrew Ireson, Sari Kovats, Kumar Mojumder, Amirul Khusru, Atiq Rahman, and Paolo Vineisl. 2011. "Drinking Water Salinity and Maternal Health in Coastal Bangladesh: Implications of Climate Change." *Environmental Health Perspectives* 119 (9): 1328–1332.

Kim, So Young, and Yael Wolinsky-Nahmias. 2014. "Cross-National Public Opinion on Climate Change: The Effects of Affluence and Vulnerability." *Global Environmental Politics* 14 (1): 79–106.

Kimball, Gayle. 2019. "Media Empowers Brave Girls to Be Global Activists." *Journal of International Women's Studies* 20 (7): 35–56.

Kinnvall, Catarina, and Helle Rydstrom, eds. 2019. *Climate Hazards, Disasters, and Gender Ramifications*. New York: Routledge.

Krange, Olve, Bjørn P. Kaltenborn, and Martin Hultman. 2019. "Cool Dudes in Norway: Climate Change Denial among Conservative Norwegian Men." *Environmental Sociology* 5 (1): 1–11.

Kvale, Steinar, and Svend Brinkmann. 2009. *Interviews: Learning the Craft of Qualitative Research Interviewing*, 2nd ed. Thousand Oaks, CA: Sage.

Lane, Ruth, and Rebecca McNaught. 2009. "Building Gendered Approaches to Adaptation in the Pacific." *Gender and Development* 17 (1): 67–80.

Leichenko, Robin. 2014. "Climate Change and Poverty: Vulnerability, Impacts, and Alleviation Strategies." *WIREs Climate Change* 5 (4): 539–556.

Leiserowitz, Anthony, Edward Maibach, Seth Rosenthal, John Kotcher, Matthew Ballew, Matthew Goldberg, and Abel Gustafson. 2018. "Climate Change in the American Mind: December 2018." New Haven, CT: Yale Program on Climate Change Communication. http://climatecommunication.yale.edu/wp-content/uploads/2019/01/Climate-Change-American-Mind-December-2018.pdf.

Levin, Kelly, Benjamin Cashore, Steven Bernstein, and Graeme Auld. 2012. "Overcoming the Tragedy of Super Wicked Problems: Constraining Our Future Selves to Ameliorate Global Climate Change." *Policy Sciences* 45 (2): 123–152.

Levine, Adeline Gordon. 1982. *Love Canal: Science, Politics, and People*. Lexington, MA: Lexington Books.

Lewis, Tania, and Emily Potter, eds. 2011. *Ethical Consumption: A Critical Introduction*. New York: Routledge.

Litfin, Karen. 2012. "Thinking Like a Planet: Gaian Politics and the Transformation of the World Food System." In *Handbook of Global Environmental Politics*, edited by Peter Dauvergne, 2nd ed., 419–430. Northampton, MA: Edward Elgar.

Littig, Beate. 2017. "Good Green Jobs for Whom? A Feminist Critique of the Green Economy." In *Routledge Handbook of Gender and Environment*, edited by Sherilyn MacGregor, 318–330. New York: Routledge.

Logsdon-Conradsen, Susan. 2011. "From Materialism to Activist Mothering: The Evolution of Mother Activism in the United States Environmental Movement." *Journal of the Motherhood Initiative* 2 (1): 9–36.

Lu, Hang, and Jonathon P. Schuldt. 2016. "Compassion for Climate Change Victims and Support for Mitigation Policy." *Journal of Environmental Psychology* 45: 192–200.

Lugones, María. 2007. "Heterosexualism and the Colonial/Modern Gender System." *Hypatia* 22 (1): 186–209.

Lugones, María. 2010. "Toward a Decolonial Feminism." *Hypatia* 25 (4): 742–759.

Lykke, Nina. 2009. "Non-Innocent Intersections of Feminism and Environmentalism." *Women, Gender and Research* 18: 36–44.

MacGregor, Sherilyn. 2006. *Beyond Mothering Earth: Ecological Citizenship and the Politics of Care*. Toronto, Canada: University of British Columbia Press.

MacGregor, Sherilyn. 2009. "A Stranger Silence Still: The Need for Feminist Social Research on Climate Change." *Sociological Review* 57 (2): 124–140.

MacGregor, Sherilyn. 2010. "'Gender and Climate Change': From Impacts to Discourses." *Journal of the Indian Ocean Region* 6 (2): 223–238.

MacGregor, Sherilyn. 2017a. "Moving beyond Impacts: More Answers to the 'Gender and Climate Change' Question." In *Understanding Climate Change through Gender Relations*, edited by Susan Buckingham and Virginie Le Masson, 15–30. New York: Routledge.

MacGregor, Sherilyn. 2017b. "Gender and Environment: An Introduction." In *Routledge Handbook of Gender and Environment*, edited by Sherilyn MacGregor, 1–24. New York: Routledge.

MacGregor, Sherilyn. 2021. "Making Matter Great Again? Ecofeminism, New Materialism and the Everyday Turn in Environmental Politics." *Environmental Politics* 30 (1–2): 41–60.

Mackenzie, Catriona, Wendy Rogers, and Susan Dodds. 2014. *Vulnerability: New Essays in Ethics and Feminist Philosophy*. Oxford: Oxford University Press.

Maclean, John. 1999. "Towards a Political Economy of Agency in Contemporary International Relations." In *Politics and Globalisation: Knowledge, Ethics, Agency*, edited by Martin Shaw, 174–201. New York: Routledge.

Magnusdottir, Gunnhildur Lily, and Annica Kronsell. 2015. "The (In)Visibility of Gender in Scandinavian Climate Policy-Making." *International Feminist Journal of Politics* 17 (2): 308–326.

Magnusdottir, Gunnhildur Lily, and Annica Kronsell, eds. 2021. *Gender, Intersectionality and Climate Institutions in Industrialised States.* New York: Routledge.

Maliniak, Daniel, Ryan Powers, and Barbara F. Walter. 2013. "The Gender Citation Gap in International Relations." *International Organization* 67 (4): 889–922.

Maniates, Michael. 2001. "Individualization: Plant a Tree, Buy a Bike, Save the World?" *Global Environmental Politics* 1 (3): 31–52.

Mann, Susan A. 2011. "Pioneers of U.S. Ecofeminism and Environmental Justice." *Feminist Formations* 23 (2): 1–25.

Manning, Susan M. 2016. "Intersectionality in Resource Extraction: A Case Study of Sexual Violence at the Porgera Mine in Papua New Guinea." *International Feminist Journal of Politics* 18 (4): 574–589.

Marino, Maria Rosaria, Marzia Romanelli, and Martino Tasso. 2013. "Dynamic Labor Supply with Taxes: The Case of Italian Couples." https://www.bancaditalia.it/pubblicazioni/altri-atti-convegni/2012-fiscal-policy-growth/Marino_Romanelli_Tasso. pdf?language_id=1.

Markham, William T. 2008. *Environmental Organizations in Modern Germany: Hardy Survivors in the Twentieth Century and Beyond.* New York: Berghahn Books.

Marsh, David, Selen A. Ercan, and Paul Furlong. 2018. "A Skin Not a Sweater: Ontology and Epistemology in Political Science." In *Theory and Methods in Political Science*, edited by Vivien Lowndes, David Marsh, and Gerry Stoker, 177–198. New York: Palgrave.

Matthew, Patricia A., ed. 2016. *Written/Unwritten: Diversity and the Hidden Truths of Tenure.* Chapel Hill: University of North Carolina Press.

Matthies, Ellen, S. Selge, and C. Klöckner. 2012. "The Role of Parental Behaviour for the Development of Behavior-Specific Environmental Norms—the Example of Recycling and Re-Use Behavior." *Journal of Environmental Psychology* 32 (3): 277–284.

Matthies, Ellen, and Hannah Wallis. 2015. "Family Socialization and Sustainable Consumption." In *Handbook of Research on Sustainable Consumption*, edited by L.A. Reisch and J. Thøgersen, 268–281. Edward Elgar.

Mayes, Eve, and Michael Everitt Hartup. 2021. "News Coverage of the School Strike for Climate Movement in Australia: The Politics of Representing Young Strikers' Emotions." *Journal of Youth Studies*. doi:10.1080/13676261.2021.1929887.

McCright, Aaron M. 2010. "The Effects of Gender on Climate Change Knowledge and Concern in the American Public." *Population and Environment* 32 (1): 66–87.

McCright, Aaron M., and Riley E. Dunlap. 2011. "Cool Dudes: The Denial of Climate Change among Conservative White Males in the United States." *Global Environmental Change* 21 (4): 1163–1172.

McCright, Aaron M., Riley E. Dunlap, and Sandra T. Marquart-Pyatt. 2016. "Political Ideology and Views about Climate Change in the European Union." *Environmental Politics* 25 (2): 338–358.

McCullagh, Elizabeth, Dare Henry-Moss, Wendy Bohon, Whitney Phinney, Theresa Jedd, and Gretchen Goldman. 2019. "Do Science. Pump. Repeat. How to Improve Scientist-Moms' Breastfeeding Experience." *Scientific American* (blog), March 18. https://blogs.scientificamerican.com/voices/do-science-pump-repeat/.

McDonald, Zack. 2018. "Surge of Domestic Violence Grows in Hurricane Michael's Wake." *Panama City News Herald*, October 24. https://www.newsherald.com/news/20181024/surge-of-domestic-violence-grows-in-hurricane-michaels-wake.

McStay, Jan R., and Riley E. Dunlap. 1983. "Male–Female Differences in Concern for the Environmental Quality." *International Journal of Women's Studies* 6 (4): 291–301.

Meijers, Marijn H. C., and Bastiaan T. Rutjens. 2014. "Affirming Belief in Scientific Progress Reduces Environmentally Friendly Behaviour." *European Journal of Social Psychology* 44 (5): 487–495.

Mellor, Mary. 1992. "Green Politics: Ecofeminist, Ecofeminine or Ecomasculine?" *Environmental Politics* 1 (2): 229–251.

Mellor, Mary. 2006. "Feminism and Environmental Ethics: A Materialist Perspective." *Thinking and Knowing About the Environment and Nature*, edited by Jules Pretty, 96–112. London: Sage.

Merchant, Carolyn. 1995. *Earthcare: Women and the Environment.* New York: Routledge.

Meyer, Ryan. 2011. "The Public Values Failures of Climate Science in the US." *Minerva* 49: 47–70.

Mies, Maria, and Vandana Shiva. 1993. *Ecofeminism.* Halifax, NS: Fernwood Publications.

Milliken, Jennifer. 1999. "The Study of Discourse in International Relations: A Critique of Research and Methods." *European Journal of International Relations* 5 (2): 225–254.

Mitchell, Daniel, Clare Heaviside, Sotiris Vardoulakis, Chris Huntingford, Giacomo Masato, Benoit P. Guillod, Peter Frumhoff, Andy Bowery, David Wallom, and Myles Allen. 2016. "Attributing Human Mortality during Extreme Heat Waves to Anthropogenic Climate Change." *Environmental Research Letters* 11 (7): 1–8.

Mitchell, Gordon, and Danny Dorling. 2003. "An Environmental Justice Analysis of British Air Quality." *Environment and Planning A* 35 (5): 909–929.

Mitchell, Ronald B., and Charli Carpenter. 2019. "Norms for the Earth: Changing the Climate on 'Climate Change.'" *Journal of Global Security Studies* 4 (4): 413–429.

Mitchell, Sara, Samantha Lange, and Holly Brus. 2013. "Gendered Citation Patterns in International Relations Journals." *International Studies Perspectives* 14 (4): 485-492.

Mohai, Paul. 1997. "Gender Differences in the Perceptions of Most Important Environmental Problems." *Race, Gender, Class* 5 (1): 153–169.

Mohanty, Chandra Talpade. 2003a. *Feminism without Borders: Decolonizing Theory, Practicing Solidarity*. Durham, NC: Duke University Press.

Mohanty, Chandra Talpade. 2003b. "'Under Western Eyes' Revisited: Feminist Solidarity through Anticapitalist Struggles." *Signs* 28 (2): 499–535.

Moi, Toril. 2001. *What Is a Woman and Other Essays.* Oxford: Oxford University Press.

Montañez, Amanda. 2018. "This Is What the Race Gap in Academia Looks Like." *Scientific American* (blog), February 19. https://blogs.scientificamerican.com/sa-visual/this-is-what-the-race-gap-in-academia-looks-like/.

Moore, Suzanne. 2019. "Greta Thunberg's Defiance Upsets the Patriarchy—and It's Wonderful." *The Guardian*, October 1. https://www.theguardian.com/commentisfree/2019/oct/01/greta-thunbergs-defiance-upsets-the-patriarchy-and-its-wonderful.

Moosa, Christina Shaheen, and Nancy Tuana. 2014. "Mapping a Research Agenda Concerning Gender and Climate Change: A Review of the Literature." *Hypatia* 29 (3): 677–694.

Moriggi, Angela. 2017. "Investigating the Gender Inequality and Climate Change Nexus in China." In *Understanding Climate Change through Gender Relations*, edited by Susan Buckingham and Virginie Le Masson, 157–172. New York: Routledge.

Naess, Lars Otto. 2013. "The Role of Local Knowledge in Adaptation to Climate Change." *WIREs Climate Change* 4 (2): 99–106.

Nagel, Joane. 2012. "Intersecting Identities and Global Climate Change." *Identities: Global Studies in Culture and Power* 19 (4): 467–476.

Nagel, Joane. 2016. *Gender and Climate Change: Impacts, Science, Policy*. New York: Routledge.

Nash, Roderick. 2001. *Wilderness and the American Mind.* New Haven, CT: Yale University Press.

Nelson, Jennifer. 2003. *Women of Color and the Reproductive Rights Movement.* New York: New York University Press.

Nelson, Julie. 2007. "Economists, Value Judgments, and Climate Change: A View from Feminist Economics." *Ecological Economics* 65 (3): 441–447.

Neumayer, Eric, and Thomas Plümper. 2007. "The Gendered Nature of Natural Disasters: The Impact of Catastrophic Events on the Gender Gap in Life Expectancy, 1981–2002." *Annals of the Association of American Geographers* 97 (3): 551–566.

Neville, Kate J., and Matthew Hoffmann. 2018. "Global Environmental Politics Research in a Time of Crisis." In *A Research Agenda for Global Environmental Politics*, edited by Peter Dauvergne and Justin Alger, 177–189. Northampton: Edward Elgar Publishing.

Newell, Peter. 2000. *Climate for Change: Non-State Actors and the Global Politics of the Greenhouse.* Cambridge: Cambridge University Press.

Ngo, Nicole S., and Radley M. Horton. 2016. "Climate Change and Fetal Health: The Impacts of Exposure to Extreme Temperatures in New York City." *Environmental Research* 44 (A): 158–164.

Nhamo, Godwell, and Senia Nhamo. 2017. "Gender and Geographical Balance: With a Focus on the UN Secretariat and the Intergovernmental Panel on Climate Change." *Gender Questions* 5 (1): 1–21.

Norgaard, Kari Marie. 2011. *Living in Denial: Climate Change, Emotions, and Everyday Life.* Cambridge, MA: MIT Press.

Norgaard, Kari, and Richard York. 2005. "Gender Equality and State Environmentalism." *Gender and Society* 19 (4): 506–522.

Obokata, Reiko, Luisa Veronis, and Robert McLeman. 2014. "Empirical Research on International Environmental Migration: A Systematic Review." *Population and Environment* 36 (1): 111–135.

Obradovich, Nick, Robyn Migliorini, Martin P. Paulus, and Iyad Rahwan. 2018. "Empirical Evidence of Mental Health Risks Posed by Climate Change." *Proceedings of the National Academy of Sciences of the United States of America* 115 (43): 10953–10958.

O'Brien, Karen. 2015. "Political Agency: The Key to Tackling Climate Change." *Science* 350 (6265): 1170–1171.

Ogburn, Stephanie Paige. 2014. "Scientist Targets of Climate Change Hate Mail Rally for Support." *Scientific American*, January 22. https://www.scientificamerican.com/article/scientist-targets-of-climate-change-hate-mail-rally-for-support/.

Ojeda, Diana, Jade S. Sasser, and Elizabeth Lunstrum. 2020. "Malthus's Specter and the Anthropocene." *Gender, Place & Culture* 27 (3): 316–332.

O'Neill, Kate, Jörg Balsiger, and Stacy D. VanDeveer. 2004. "Actors, Norms, and Impact: Recent International Cooperation Theory and the Influence of the Agent-Structure Debate." *Annual Review of Political Science* 14 (7): 149–175.

O'Neill, Kate, Erika Weinthal, Kimberly R. Marion Suiseeya, Steven Bernstein, Avery Cohn, Michael W. Stone, and Benjamin Cashore. 2013. "Methods and Global Environmental Governance." *Annual Review of Environment and Resources* 38: 441–471.

Oven, Katie, Jonathan Rigg, Shubheksha Rana, Arya Gautam, and Toran Singh. 2019. "#leave noonebehind: Women, Gender Planning and Disaster Risk Reduction in Nepal." In *Climate Hazards, Disasters, and Gender Ramifications*, edited by Catarina Kinnvall and Helle Rydstrom, 138–164. New York: Routledge.

Paavola, Jouni. 2006. "Justice in Adaptation to Climate Change in Tanzania." In *Fairness in Adaptation to Climate Change*, edited by W. Neil Adger, Jouni Paavola, Saleemul Huq, and M. J. Mace, 201–222. Cambridge, MA: MIT Press.

Pacifici, Michela, Wendy B. Foden, Piero Visconti, James E. M. Watson, Stuart H. M. Butchart, Kit M. Kovacs, Brett R. Scheffers, et al. 2015. "Assessing Species Vulnerability to Climate Change." *Nature Climate Change* 5 (3): 215–224.

Panitchpakdi, Supachai. 2008. "Secretary-General of UNCTAD, Statement to Third Global Congress of Women in Politics and Governance: Gender in Climate Change and Disaster Risk Reduction." Makati, The Philippines: United Nations Conference on Trade and Development. http://unctad.org/en/pages/SGStatementArchive.aspx?ReferenceItemId=16455.

Partzsch, Lena. 2017. "'Power with' and 'Power to' in Environmental Politics and the Transition to Sustainability." *Environmental Politics* 26 (2): 193–211.

Pasgaard, Maya, Bo Dalsgaard, Pietro K. Maruyama, Brody Sandel, and Niels Strange. 2015. "Geographical Imbalances and Divides in the Scientific Production of Climate Change Knowledge." *Global Environmental Change* 35: 279–288.

Pasgaard, Maya, and Niels Strange. 2013. "A Quantitative Analysis of the Causes of the Global Climate Change Research Distribution." *Global Environmental Change* 23 (6): 1684–1693.

Pellow, David N. 2016. "Toward a Critical Environmental Justice Studies: Black Lives Matter as an Environmental Justice Challenge." *Du Bois Review* 13 (2): 1–16. https://doi.org/http://dx.doi.org/10.1017/S1742058X1600014X.

Pepermans, Yves, and Pieter Maeseele. 2016. "The Politicization of Climate Change: Problem or Solution?" *WIRES Climate Change* 16 (4): 478–485.

Perkins, Tracy E. 2012. "Women's Pathways into Activism: Rethinking the Women's Environmental Justice Narrative in California's San Joaquin Valley." *Organization and Environment* 25 (1): 76–94.

Pettenger, Mary E., ed. 2007. *The Social Construction of Climate Change: Power, Knowledge, Norms, Discourses*, New York: Routledge.

Pierrat, Chantal. 2015. "8 Badass Environmentalists You Should Know." *HuffPost* (blog), April 22. https://www.huffpost.com/entry/8-badass-environmentalists-you-should-know_b_7119676.

Plumwood, Val. 2002. *Environmental Culture: The Ecological Crisis of Reason*. New York: Routledge.

Plumwood, Val. 2006. "Feminism." In *Political Theory and the Ecological Challenge*, edited by Andrew Dobson and Robyn Eckersley, 51–74. New York: Cambridge University Press.

Poushter, Jacob, and Christine Huang. 2019. "Climate Change Still Seen as the Top Global Threat, but Cyberattacks a Rising Concern." Pew Research Center. *https://www.pewresearch.org/global/2019/02/10/climate-change-still-seen-as-the-top-global-threat-but-cyberattacks-a-rising-concern/.*

Priestley, Mark, and Laura Hemingway. 2007. "Disability and Disaster Recovery." *Journal of Social Work in Disability & Rehabilitation* 5 (3): 23–42.

Princen, Thomas. 2002. "Consumption and Its Externalities: Where Economy Meets Ecology." In *Confronting Consumption*, edited by Thomas Princen, Michael Maniates, and Ken Conca, 23–42. Cambridge, MA: MIT Press.

Princen, Thomas. 2009. "Long-Term Decision-Making: Biological and Psychological." *Global Environmental Politics* 9 (3): 9–19.

Pulé, Paul, and Martin Hultman. 2019. "Industrial/Breadwinner Masculinities: Understanding the Complexities of Climate Change Denial." In *Climate Hazards, Disasters, and Gender Ramifications*, edited by Catarina Kinnvall and Helle Rydstrom, 86–100. New York: Routledge.

Pulé, Paul M., and Martin Hultman, eds. 2021. *Men, Masculinities, and Earth: Contending with the (m)Anthropocene*. New York: Palgrave Macmillan.

Puri, Lakshmi. 2015. "Women and Girls Are Essential Climate Actors." *UN Women*, December 21. http://www.unwomen.org/en/news/stories/2015/12/lakshmi-puri-at-globe-internationals-annual-legislators-summit#sthash.cPitfdcU.dpuf.

Raney, Tracey, and Mackenzie Gregory. 2019. "Green with Rage: Women Climate Change Leaders Face Online Attacks." *The Conversation*, September 12. http://theconversation.com/green-with-rage-women-climate-change-leaders-face-online-attacks-123155.

Rao, Namratha, and Anita Raj. 2019. "Women May Be More Vulnerable to Climate Change but Data Absent." IndiaSpend. https://www.indiaspend.com/women-may-be-more-vulnerable-to-climate-change-but-data-absent.

Rao, Rahul. 2014. "Queer Questions." *International Feminist Journal of Politics* 16 (2): 199–217.

Rao, Smitha. 2020. "A Natural Disaster and Intimate Partner Violence: Evidence over Time." *Social Science and Medicine* 247: 112804.

Reyes-García, Victoria, Álvaro Fernández-Llamazares, Maximilien Guèze, Ariadna Garcés, Miguel Mallo, Margarita Vila-Gómez, and Marina Vilaseca. 2016. "Local Indicators of Climate Change: The Potential Contribution of Local Knowledge to Climate Research." *WIREs Climate Change* 7 (1): 109–124.

Ribot, Jesse. 2010. "Vulnerability Does Not Fall from the Sky: Toward Multiscale, Pro-Poor Climate Policy." In *Social Dimensions of Climate Change: Equity and Vulnerability in a Warming World*, edited by Robin Mearns and Andrew Norton, 47–74. Washington, DC: The World Bank.

Rittel, Horst W. J., and Melvin M. Webber. 1973. "Dilemmas in a General Theory of Planning." *Policy Sciences* 4 (2): 155–169.

Robinson, Fiona. 2011. *The Ethics of Care: A Feminist Approach to Human Security*. Philadelphia: Temple University Press.

Rocheleau, Dianne, Barbara Thomas-Slayter, and Esther Wangari, eds. 1996. *Feminist Political Ecology: Global Issues and Local Experiences*. New York: Routledge.

Rogers, Stephanie. 2010. "The 20 Most Influential Women in Green." *Ecosalon* (blog), December 7. http://ecosalon.com/the-19-most-influential-women-in-green.

Rosen, Rachel, and Katherine Twamley. 2018. "Introduction: The Woman–Child Question: A Dialogue in the Borderlands." In *Feminism and the Politics of Childhood: Friends or Foes?*, edited by Rachel Rosen and Katherine Twamley, 1–20. London: University College London Press.

Rydstrom, Helle. 2019. "Disasters, Ruins, and Crisis: Masculinity and the Ramifications of Storms in Vietnam." *Ethnos* 85 (2): 351–370. https://doi.org/10.1080/00141844.2018.1561490.

Rylander, Charlotta, Jon Øyvind Odland, and Torkjel Manning Sandanger. 2013. "Climate Change and the Potential Effects on Maternal and Pregnancy Outcomes: An Assessment of the Most Vulnerable—the Mother, Fetus, and Newborn Child." *Global Health Action* 6 (1): 19538–19547.

Salleh, Ariel. 1997. *Ecofeminism as Politics: Nature, Marx and the Postmodern*. New York: Zed Books.

Sandilands, Catriona. 1999. *The Good-Natured Feminist: Ecofeminism and the Quest for Democracy*. Minneapolis: University of Minnesota Press.

Sapra, Sonalini. 2009. "Participatory Democracy and Social Justice: The Politics of Women's Environmental Action in India." PhD dissertation, Vanderbilt University.

Sasser, Jade S. 2018. *On Infertile Ground: Population Control and Women's Rights in the Era of Climate Change*. New York: New York University Press.

Scerri, Andy, and Liam Magee. 2012. "Green Householders, Stakeholder Citizenship and Sustainability." *Environmental Politics* 21 (3): 387–411.

Schaeffer, Katherine. 2019. "Among U.S. Couples, Women Do More Cooking and Grocery Shopping than Men." Pew Research Center. https://www.pewresearch.org/fact-tank/2019/09/24/among-u-s-couples-women-do-more-cooking-and-grocery-shopping-than-men.

Schlosberg, David, and David Carruthers. 2010. "Indigenous Struggles, Environmental Justice, and Community Capabilities." *Global Environmental Politics* 10 (4): 12–35.

Schreurs, Miranda A. 2004. *Environmental Politics in Japan, Germany, and the United States*. New York: Cambridge University Press.

Schumann, Karina, and Michael Ross. 2010. "Why Women Apologize More than Men: Gender Differences in Thresholds for Perceiving Offensive Behavior." *Psychological Science* 21 (11): 1649–1655.

Scott, Daniel, Michael Hall, and Stefan Gössling. 2019. "Global Tourism Vulnerability to Climate Change." *Annals of Tourism Research* 77: 49–61.

Seager, Joni. 1996. "'Hysterical Housewives' and Other Mad Women: Grassroots Environmental Organizing in the United States." In *Feminist Political Ecology*, edited by Dianne Rocheleau, Barbara Thomas-Slayter, Esther Wangari, 271–283. New York: Routledge.

Seager, Joni. 1999. "Patriarchal Vandalism: Militaries and the Environment." In *Dangerous Intersections: Feminist Perspectives on Population, Environment, and Development*, edited by Jael Silliman and Ynestra King, 163–188. Cambridge, MA: South End Press.

Seager, Joni. 2003. "Pepperoni or Broccoli? On the Cutting Wedge of Feminist Environmentalism." *Gender, Place and Culture* 10 (2): 167–174.

Seager, Joni. 2009. "Death by Degrees: Taking a Feminist Hard Look at the 2° Climate Policy." *Women, Gender and Research* 18: 11–21.

Searle, Kristina, and Kathryn Gow. 2010. "Do Concerns about Climate Change Lead to Distress?" *International Journal of Climate Change Strategies and Management* 2 (4): 362–379.

Segal, Brenda, and Jeffrey S. Podoshen. 2013. "An Examination of Materialism, Conspicuous Consumption and Gender Differences." *International Journal of Consumer Studies* 37 (2): 189–198.

Seymour, Nicole. 2013. *Strange Natures: Futurity, Empathy, and the Queer Ecological Imagination*. Champaign: University of Illinois Press.

Shearer, Christine. 2012. "The Social Construction of Alaska Native Vulnerability to Climate Change." *Race, Gender and Class* 19 (1–2): 61–79.

Sheffield, Perry E., and Philip J. Landrigan. 2011. "Global Climate Change and Children's Health: Threats and Strategies for Prevention." *Environmental Health Perspectives* 119 (3): 291–298.

Shepherd, Marshall, and Binita K. C. 2015. "Climate Change and African Americans in the USA." *Geography Compass* 9 (11): 579–591.

Shiva, Vandana. 1989. *Staying Alive: Women, Ecology, and Development.* Atlantic Heights, NJ: Zed Books.

Sjoberg, Laura, ed. 2010. *Gender and International Security: Feminist Perspectives.* Routledge Critical Security Studies Series. New York: Routledge.

Smith, N., and A. Leiserowitz. 2013. "American Evangelicals and Global Warming." *Global Environmental Change* 23 (5): 1009–1017.

Soper, Kate. 2009. "Beyond Consumerism: Reflections on Gender Politics, Pleasure and Sustainable Consumption." *Women, Gender and Research* 18 (3–4): 92–100.

Staddon, Sam C., Andrea Nightingale, and Shyam K. Shrestha. 2014. "The Social Nature of Participatory Ecological Monitoring." *Society & Natural Resources* 27 (9): 899–914.

Staudt, Kathleen, Tony Payan, and Z. Anthony Kruszewski, eds. 2009. *Human Rights along the U.S.-Mexico Border: Gendered Violence and Insecurity.* Tucson: University of Arizona Press.

Stern, Maria. 2006. "Racism, Sexism, Classism, and Much More: Reading Security-Identity in Marginalized Sites." In *Feminist Methodologies for International Relations*, edited by Brooke A. Ackerly, Maria Stern, and Jacqui True, 174–198. New York: Cambridge University Press.

Stern, Paul C., Thomas Dietz, and Linda Kalof. 1993. "Value Orientations, Gender, and Environmental Concern." *Environmental Behavior* 25 (5): 322–348.

Stolle, Dietlind, and Michele Micheletti. 2005. "The Gender Gap Reversed: Political Consumerism as a Women-Friendly Form of Civic and Political Engagement." In *Gender and Social Capital*, edited by Elisabeth Gidengil and Brenda O'Neill, 45–72. New York: Routledge.

Stover, Kayla, and Sherry Cable. 2017. "American Women's Environmental Activism: Motivations, Experiences, and Transformations." In *The Oxford Handbook of U.S. Women's Social Movement Activism*, edited by Holly J. McCammon, Verta Taylor, Jo Reger, and Rachel L. Einwohner, 685–707. New York: Oxford University Press.

Sturgeon, Noël. 1997. *Ecofeminist Natures: Race, Gender, Feminist Theory and Political Action.* New York: Routledge.

Sturgeon, Noël. 2008. *Environmentalism in Popular Culture: Gender, Race, Sexuality, and the Politics of the Natural.* Tucson: University of Arizona Press.

Styles, John, and Amanda Vickery. 2007. *Gender, Taste, and Material Culture in Britain and North America, 1700–1830.* New Haven, CT: Yale University Press.

Sultana, Farhana. 2009. "Fluid Lives: Subjectivities, Gender and Water in Rural Bangladesh." *Gender, Place and Culture* 16 (4): 427–444.

Sultana, Farhana. 2011. "Suffering for Water, Suffering from Water: Emotional Geographies of Resource Access, Control and Conflict." *Geoforum* 42 (2): 163–172.

Sundström, Aksel, and Aaron M. McCright. 2014. "Gender Differences in Environmental Concern among Swedish Citizens and Politicians." *Environmental Politics.* 23 (6): 1082–1095.

Sutherland, Nikki, Paul Bolton, Steven Kennedy, and Suzanna Hinson. 2018. "Fuel Poverty." United Kingdom Parliament, House of Commons Library, December 11. https://researchbriefings.parliament.uk/ResearchBriefing/Summary/CDP-2018-0276.

Swaminathan, Nikhil. 2017. "The Unsustainable Whiteness of Green." *Grist* (blog). https://grist.org/feature/the-unsustainable-whiteness-of-green.

Swim, Janet K., Ashley J. Gillis, and Kaitlynn J. Hamaty. 2020. "Gender Bending and Gender Conformity: The Social Consequences of Engaging in Feminine and Masculine Pro-Environmental Behaviors." *Sex Roles* 82 (5–6): 363–385. https://doi.org/https://doi.org/10.1007/s11199-019-01061-9.

Sze, Julie. 2017. "Gender and Environmental Justice." In *Routledge Handbook of Gender and Environment*, edited by Sherilyn MacGregor, 159–168. New York: Routledge.

Taylor, Dorceta E. 2014. "The State of Diversity in Environmental Organizations: Mainstream NGOs, Foundations, Government Agencies." Green 2.0. https://vaipl.org/wp-content/uploads/2014/10/ExecutiveSummary-Diverse-Green.pdf.

Taylor, Dorceta E. 2016. *The Rise of the American Conservation Movement, Privilege, and Environmental Protection.* Durham, NC: Duke University Press.

Taylor, Dorceta E. 2018. "Diversity in Environmental Organizations: Reporting and Transparency." Report No. 1. Ann Arbor: University of Michigan, School for Environment and Sustainability.

Team, Victoria, and Eyob Hassen. 2016. "Climate Change and Complexity of Gender Issues in Ethiopia." In *Systemic Crises of Global Climate Change: Intersections of Race, Class and Gender*, edited by Phoebe Godfrey and Denise Torres, 314–326. New York: Routledge.

Terry, Geraldine. 2009. "No Climate Justice without Gender Justice: An Overview of the Issues." *Gender and Development* 17 (1): 5–18.

Thew, Harriet. 2018. "Youth Participation and Agency in the United Nations Framework Convention on Climate Change." *International Environmental Agreements* 18: 369–389.

Thomas, Kimberley Anh, and Benjamin P. Warner. 2019. "Weaponizing Vulnerability to Climate Change." *Global Environmental Change* 57: 101928. https://doi.org/https://doi.org/10.1016/j.gloenvcha.2019.101928.

Tickner, J. Ann. 1997. "You Just Don't Understand: Troubled Engagements between Feminists and IR Theorists." *International Studies Quarterly* 41 (4): 611–632.

Tickner, J. Ann. 2014. *A Feminist Voyage through International Relations.* New York: Oxford University Press.

Tidwell, Christy, and Bridgitte Barclay. 2019. *Gender and Environment in Science Fiction.* Lanham, MD: Lexington Books.

Tobin-Gurley, Jennifer, Robin Cox, Lori Peek, Kylie Pybus, Dmitriy Maslenitsyn, and Cheryl Heykoop. 2016. "Youth Creating Disaster Recovery and Resilience in Canada and the United States." In *Men, Masculinities and Disaster*, edited by Elaine Enarson and Bob Pease, 152–161. New York: Routledge.

Tonkiss, Fran. 2004. "Analysing Text and Speech: Content and Discourse Analysis." In *Researching Society and Culture*, edited by Clive Seale, 367–382. Thousand Oaks, CA: Sage.

Toomey, Diane. 2018. "How Green Groups Became So White and What to Do about It." Yale Environment 360. https://e360.yale.edu/features/how-green-groups-became-so-white-and-what-to-do-about-it.

Townsend, Solitaire. 2018. "46 Sustainability Leaders (Who Are Also Women)." *Forbes* (blog), July 18. https://www.forbes.com/sites/solitairetownsend/2018/07/18/46-sustainability-leaders-who-are-also-women/#3ca5ccf51110.

Tran, Dalena, Joan Martínez-Alier, Grettel Navas, and Sara Mingorría. 2020. "Gendered Geographies of Violence: A Multiple Case Study Analysis of Murdered Women Environmental Defenders." *Journal of Political Ecology* 27 (1): 1189–1212.

Tronto, Joan. 2013. *Caring Democracy: Markets, Equality and Justice.* New York: New York University Press Press.

True, Jacqui. 2003. *Gender Globalization and Postsocialism: The Czech Republic after Communism.* New York: Columbia University Press.

True, Jacqui. 2012. *The Political Economy of Violence against Women*. New York: Oxford University Press.

Tuana, Nancy. 2013. "Gendering Climate Knowledge for Justice: Catalyzing a New Research Agenda." In *Research, Action, and Policy: Addressing the Gendered Impacts of Climate Change*, edited by Margaret Alston and Kerri Whittenbury, 17–31. New York: Springer.

Turner, Bryan S. 2006. *Vulnerability and Human Rights.* University Park: Pennsylvania State University Press.

Tuvel, Rebecca. 2015. "Sourcing Women's Ecological Knowledge: The Worry of Epistemic Objectification." *Hypatia* 30 (2): 319–336.

Underdal, Arild. 2017. "Climate Change and International Relations (after Kyoto)." *Annual Review of Political Science* 20 (1): 169–188.

UNCED. *See* United Nations Commission on Environment and Development.

UNDP. *See* United Nations Development Programme.

UNFCCC. *See* United Nations Framework Convention on Climate Change.

United Nations. 2015. "Transforming Our World: The 2030 Agenda for Sustainable Development." https://sustainabledevelopment.un.org/sdgs.

United Nations Climate Change. 2019. "Christiana Figueres." https://unfccc.int/about-us/the-executive-secretary/former-executive-secretary-ms-christiana-figueres.

United Nations Commission on Environment and Development. 1992. *Agenda 21*. http://sustainabledevelopment.un.org/content/documents/Agenda21.pdf.

United Nations Development Programme. 2015. "Human Development Report 2015: Work for Human Development." http://hdr.undp.org/sites/default/files/2015_human_development_report.pdf.

United Nations Environment Programme. 2019. "Promoting Gender Equality and the Human Rights and Empowerment of Women and Girls in Environmental Governance." https://wedocs.unep.org/bitstream/handle/20.500.11822/28481/English.pdf?sequence=3&isAllowed=y.

United Nations Framework Convention on Climate Change. 2001. "Report of the Conference of the Parties on Its Seventh Session, Held at Marrakesh from 29 October to 10 November 2001." https://unfccc.int/resource/docs/cop7/13a04.pdf.

United Nations Framework Convention on Climate Change. 2015. "The Paris Agreement." http://unfccc.int/paris_agreement/items/9485.php.

United Nations Framework Convention on Climate Change. 2019. "Gender Composition: Report by the Secretariat." FCCC/CP/2019/9. https://unfccc.int/sites/default/files/resource/CP2019_09E.pdf.

Unsworth, Kerrie L., and Kelly S. Fielding. 2014. "It's Political: How the Salience of One's Political Identity Changes Climate Change Beliefs and Policy Support." *Global Environmental Change* 27: 131–137.

VanLeuvan, Patricia. 2004. "Young Women's Science/Mathematics Career Goals from Seventh Grade to High School Graduation." *Journal of Educational Research* 97 (5): 248–267.

Vickery, Amanda. 2006. "His and Hers: Gender, Consumption and Household Accounting in Eighteenth-Century England." *Past and Present* 1 (1): 12–38.

Vu, Hong Tien, Yuchen Liu, and Duc Vinh Tran. 2019. "Nationalizing a Global Phenomenon: A Study of How the Press in 45 Countries and Territories Portrays Climate Change." *Global Environmental Change* 58: 101942. https://doi.org/10.1016/j.gloenvcha.2019.101942.

Waldman, Scott, and Niina Heikkinen. 2018. "As Climate Scientists Speak Out, Sexist Attacks Are on the Rise." *Scientific American*, August 22. https://www.scientificamerican.com/article/as-climate-scientists-speak-out-sexist-attacks-are-on-the-rise.

Walfisch, Tamar, Dina Van Dijk, and Ronit Kark. 2013. "Do You Really Expect Me to Apologize? The Impact of Status and Gender on the Effectiveness of an Apology in the Workplace." *Journal of Applied Social Psychology* 43 (7): 1446–1458.

Wang, Susie, Zoe Leviston, Mark Hurlstone, Carmen Lawrence, and Iain Walker. 2018. "Emotions Predict Policy Support: Why It Matters How People Feel about Climate Change." *Global Environmental Change* 50: 25–40.

Wapner, Paul. 2012. "After Nature: Environmental Politics in a Postmodern Age." In *Handbook of Global Environmental Politics*, edited by Peter Dauvergne, 2nd edition, 431–442. Northampton, MA: Edward Elgar.

Warner, Benjamin P., and Christopher Kuzdas. 2016. "Manufactured Global-Change Risk Pathways in Industrial-Based Agrarian Development." *Climate and Development* 8 (5): 385–396.

Warren, Karen J. 1997. *Ecofeminism: Women, Culture, Nature.* Bloomington: Indiana University Press.

Watts, Nick. 2015. "Health and Climate Change: Policy Responses to Protect Health." *The Lancet* 386 (10006): 1861–1914.

Watts, Nick, Markus Amann, Sonja Ayeb-Karlsson, Kristine Belesova, Timothy Bouley, Maxwell Boykoff, Peter Byass, et al. 2018. "The Lancet Countdown on Health and Climate Change: From 25 Years of Inaction to a Global Transformation for public Health." *The Lancet* 391 (10120): 581–630.

Waylen, Georgina. 1997. "Gender, Feminism and Political Economy." *New Political Economy* 2 (2): 205–220.

Weart, Spencer. 2011. "The Development of the Concept of Dangerous Anthropogenic Climate Change." In *The Oxford Handbook of Climate Change and Society*, edited by John S. Dryzek, Richard B. Norgaard, and David Schlosberg, 67–81. New York: Oxford University Press.

Whetung, Madeline. 2019. "(En)Gendering Shoreline Law: Nishnaabeg Relational Politics along the Trent Severn Waterway." *Global Environmental Politics* 19 (3): 16–32.

Whittenbury, Kerri. 2012. "Climate Change, Women's Health, Wellbeing and Experiences of Gender Based Violence in Australia." In *Research, Action and Policy: Addressing the Gendered Impacts of Climate Change*, edited by Margaret Alston and Kerri Whittenbury, 207–221. New York: Springer.

Whyte, Kyle Powys, and Chris Cuomo. 2016. "Ethics of Caring in Environmental Ethics: Indigenous and Feminist Philosophies." In *The Oxford Handbook of Environmental Ethics*, edited by Stephen M. Gardiner and Allen Thompson, 234–247. Oxford: Oxford University Press.

Wibben, Annick, ed. 2016. *Researching War: Feminist Methods, Ethics and Politics*, New York: Routledge.

Wibben, Annick T. R. 2011. *Feminist Security Studies: A Narrative Approach.* New York: Routledge.

Wilson, Joanna, and Eric Chu. 2020. "The Embodied Politics of Climate Change: Analysing the Gendered Division of Environmental Labour in the UK." *Environmental Politics* 29 (6): 1085–1104.

Wisner, Ben, Piers Blaikie, Terry Cannon, and Ian Davis. 2003. *At Risk: Natural Hazards, People's Vulnerability and Disasters.* 2nd edition. New York: Routledge.

Wolbring, Gregor, and Verlyn Leopatra. 2012. "Climate Change, Water, Sanitation and Energy Insecurity: Invisibility of People with Disabilities." *Canadian Journal of Disability Studies* 1 (3): 66–90. https://doi.org/https://doi.org/10.15353/cjds.v1i3.58.

Wolsko, Christopher, Hector Ariceaga, and Jesse Seiden. 2016. "Red, White, and Blue Enough to Be Green: Effects of Moral Framing on Climate Change Attitudes and Conservation Behaviors." *Journal of Experimental Social Psychology* 65 (July): 7–19.

Women Also Know Stuff. 2019. Web page. https://womenalsoknowstuff.com/.

World Resources Institute. 2018. "Christiana Figueres." https://www.wri.org/profile/christiana-figueres.

Wrathall, David J. 2012. "Migration amidst Social-Ecological Regime Shift: The Search for Stability in Garífuna Villages of Northern Honduras." *Human Ecology* 40 (4): 583–596.

Wroth, Katharine. 2016. "Mothers Unleash Their Organizing Power on Climate." *Grist*, December 26. https://grist.org/climate-energy/mothers-unleash-their-organizing-power-on-climate/.

Xiao, Chenyang, and Aaron McCright. 2012. "Explaining Gender Differences in Concern about Environmental Problems in the United States." *Society & Natural Resources* 25 (11): 1067–1084.

Xiao, Chenyang, and Aaron M. McCright. 2014. "A Test of the Biographical Availability Argument for Gender Differences in Environmental Behaviors." *Environment and Behavior* 46 (2): 241–263.

Yeo, Alice, Robin Legard, Jill Keegan, Kit Ward, Carol McNaughton Nicholls, and Jane Lewis. 2014. "In-Depth Interviews." In *Qualitative Research Practice*, edited by Jane Ritchie, Jane Lewis, Carol McNaughton Nicholls, and Rachel Ormston, 177–210. Washington, DC: Sage.

Yeo, Sophie. 2018. "Fighting Sexism at the Intergovernmental Panel on Climate Change." *Pacific Standard*, July 20. https://psmag.com/environment/fixing-sexism-at-the-intergovernmental-panel-on-climate-change.

Young, Iris Marion. 2002. "Lived Body vs. Gender: Reflections on Social Structure and Subjectivity." *Ratio* 15: 410–428.

Zainulbhai, Hani. 2015. "Women, More than Men, Say Climate Change Will Harm Them Personally." *Pew Research Center*. http://www.pewresearch.org/fact-tank/2015/12/02/women-more-than-men-say-climate-change-will-harm-them-personally/.

Zhou, Yuan, and Xiaoyan Sun. 2020. "Toward Gender Sensitivity: Women and Climate Change Policies in China." *International Feminist Journal of Politics* 22 (1): 127–149.

Zurawski, Karen. 2018. "Domestic Violence Increases after Harvey." *Houston Chronicle*, October 16. https://www.chron.com/neighborhood/katy/news/article/KCM-Domestic-violence-increases-after-Harvey-13308826.php.

Index

Ackerly, Brooke, 15, 16
Adaptation, 2, 3, 4, 9, 29, 33, 48, 52, 57, 60, 62–63, 75, 98, 99, 103, 148, 165, 179, 183, 201, 206
Agency, 7, 11, 21, 23, 25, 28, 57, 60, 61, 64, 70, 94, 96, 124, 134–136, 140–142, 147, 148, 155–156, 165–168, 170, 171, 172, 176, 186, 188, 189, 191, 193, 202–206
Agenda 21, 69, 83, 90, 91
Agriculture, 27, 30, 31, 33, 35, 48, 53, 54, 60, 73, 79, 99, 105, 134, 152, 157, 183, 200
Alaimo, Stacy, 195
Anshelm, Jonas, 10, 170
Arora-Jonsson, Seema, 62, 168
Auld, Graeme, 3
Australia, 26, 27

Bäckstrand, Karin, 130
Bangladesh, 25, 26, 29–30, 48, 53, 80, 141
Bernstein, Steven, 3
Brazil, 25, 33, 100
Buckingham, Susan, 32
Butler, Cameron, 70

Canada, 25, 27, 129, 170
Capitalism, 40, 88, 147, 178
Carbon dioxide, 40, 89, 141, 145, 158, 181, 184
Care/caregiving, 12, 29, 30, 35, 39, 52, 55, 56, 57, 63, 66–80, 81, 83, 86–90, 104, 122, 123, 124, 127, 145, 155, 167, 169, 181, 187, 195–197, 198
Cashore, Benjamin, 3
Chan, Jody R., 129
Chavis, Benjamin, 47
Children, 29, 30, 35, 38, 50, 51, 56, 65, 66, 68, 69, 70, 72, 74–80, 81–91, 94, 95, 103, 104, 115, 117, 122, 123, 130, 134, 155, 159, 170, 195, 197
China, 26
Chu, Eric, 179
Clapp, Jennifer, 17
Clean energy, 89, 118, 119, 133, 140, 148, 150, 157–158, 160, 165, 181, 184
Climate change
  dominating academic and policy focus, 3–4, 177
  exacerbating vulnerability, 2, 5, 10, 39, 48, 56, 60, 62, 176, 183
  and mental health, 2, 80, 83, 90
  perceptions of, 55–56, 90, 124, 165, 176–184
  politicization of, 2
  as a "super wicked problem," 3, 177
Climate denial, 10, 98, 100, 101–104, 126, 131, 170, 171, 179
Climate economics, 9, 98–99

Climate science, 4, 9, 110–114, 115, 116, 121, 122, 123, 126, 127, 131, 163, 180–181, 182, 190, 198, 199, 202
Consumption, 84, 85, 103, 116, 135–136, 140, 142–148, 156, 165, 166–168, 196, 202, 205
Cox, Robert, 16
Crenshaw, Kimberlé, 6
Cullen, Heidi, 110
Cuomo, Chris, 67
Curnow, Joe, 129
Cyclone Aila, 29, 80

Dauvergne, Peter, 17
d'Eaubonne, Françoise, 95
Demetriades, Justina, 61
Dhaliwal, Suzanne, 172
Di Chiro, Giovanna, 196
Disability, 32, 49–51, 60, 172
Discourse analysis, 19–23
Drought, 2, 31, 35, 36, 37, 52, 53, 56, 79, 149, 157
Dunlap, Riley E., 100

Ecofeminism, 7, 67, 95, 96, 142–143
Education, 30, 32, 52, 53, 65, 75, 76, 80, 84, 98, 103, 115, 119, 120, 125, 126, 139, 163, 199
Elliott, James R., 48
Enloe, Cynthia, 16, 23, 56, 70, 189
Environmental attitudes, 6, 33–34, 67, 68–69
Environmental justice, 43, 45, 47, 58–59, 87, 191–192, 201–202
Environmental knowledge, 67, 99, 104–106, 124, 187
Environmental racism, 47, 137, 194
Environmental Revolution, 40, 42–43, 44, 161
"Environmental saviors" trope, 9, 25, 62, 142, 168, 201
"Environmental victims" trope, 9, 25, 61–62, 82, 168
Epstein, Charlotte, 23
Ercan, Selen A.,18
Esplen, Emily, 61
Ethiopia, 26
Europe, 3, 52, 54, 101, 142, 144, 170, 188, 199
Evans, Gary, 69

Femininity, 5–6, 8, 9, 12, 82, 88, 92, 110, 113, 127, 143, 152–153, 163, 166, 168, 181, 184, 197, 201
Feminist constructivism, 20–21, 175
Feminist international relations, 11, 17, 18, 23, 28, 192
Feminist methodology, 15–28
Feminization of environmentalism, 143, 153, 166
Fertility, 63, 83–84
Figueres, Christiana, 97, 107–108, 110, 124, 142
Flooding, 1, 2, 29, 30, 37, 38, 46, 48, 50, 54, 56, 57, 69, 79
Food, 2, 9, 29, 30, 35, 48, 52, 56, 78, 80, 81, 83, 91, 104–105, 141, 144, 167, 186, 200
  food security, 9, 29, 30, 81, 83, 186
Ford, Lucy, 155
Fossil fuel, 40, 56, 65, 100, 101, 145, 156–157, 160, 165, 181, 182
Foster, Emma, 91
Fox Keller, Evelyn, 113
Furlong, Paul, 18

Gambia, 149
Gay-Antaki, Miriam, 130
Gender
  and climate change scholarship, 7–11
  defining, 5–7, 184–191
  and environmental attitudes, 6, 33–34, 68–69, 100–102
  inequality, 60–61
  and labor, 30, 35–37, 39, 52, 53, 57, 61, 67, 79, 80–83, 91, 92, 99, 104–105, 124, 143–144, 148, 166, 169, 198

norms, 6, 7, 9, 11, 20, 28, 32, 34, 35, 42, 56, 67, 73, 87, 89, 106, 124, 136, 143, 148, 152, 167, 170, 186, 193, 196, 197
and science, 110–114
as a social construct, 5
GenderCC, 164
Gendered violence, 6, 36–37, 38, 39, 171
Geoengineering, 160, 181, 182
Germany, 25, 27, 113, 145, 148, 153, 154, 163
Ghana, 99
Gibbs, Lois, 94, 118
Glazebrook, Trish, 99
Global environmental politics (GEP), 3, 10, 11, 15–21, 135, 143, 168, 192
Global Environmental Politics (journal), 11
Global North, 2, 3, 5, 9, 21, 25–28, 34, 36, 39, 42, 48, 52, 53–54, 55, 57, 58, 62, 63, 66, 82, 84, 91, 98, 125, 168, 171, 172, 178, 179, 187, 189, 190, 191, 192, 194, 197, 199, 200, 201, 202, 203, 205
Global South, 5, 9, 20, 25, 26, 27, 39, 48, 52–54, 55, 57, 58, 60, 62, 63, 64, 84, 90, 91, 105, 106, 124, 125, 130, 141, 149, 150, 154, 168, 172, 187, 188, 189, 191, 192, 194, 200, 202, 205, 206
Greenhouse gases, 26, 79, 83, 148, 182
Green New Deal, 194–195
Grønhøj, Alice, 69
Guatemala, 141, 149

Hajer, Maarten, 19
Hamilton, Lawrence, 68
Hayhoe, Katherine, 110
Health, 2, 34, 47, 49, 51, 62, 65, 68, 74, 78–80, 83, 84, 89, 90, 94, 95, 115, 117, 175
Heat waves, 2, 3, 79
Higgins, Maeve, 86
Hirsi, Isra, 170
Hoffman, Matthew, 18
Howell, Junia, 48
Hultman, Martin, 10, 100, 170
Hurricane Harvey, 39, 48
Hurricane Katrina, 38–39, 133
Hurricane Michael, 39
Huynh, Phuong T. A., 32

India, 39, 103, 149
Indigenous peoples, 24, 31, 58, 59, 91, 98, 103, 138, 149, 163, 172, 188, 191, 194, 199, 202
indigenous women, 149, 163, 172, 191, 194, 202
Industrial Revolution, 40–41, 110, 156
Intergovernmental Panel on Climate Change (IPCC), 2, 50, 59, 69, 83, 95, 130–131, 158, 179
and gender, 130–131
International Friends of Nature, 42
International Relations (IR), 11, 15, 16, 17, 18, 20, 23, 28, 134, 192. *See also* Feminist international relations
International Studies Quarterly, 15
International Union of Forestry Research Organizations, 42
Intersectionality, 6–7, 21, 33, 47, 58, 99–100, 121, 149, 190, 202–203, 204
Intimate partner violence (IPV), 38–39

Jaggar, Alison, 191
Jinnah, Sikina, 1–2
Johnson, Boris, 183

Kaijser, Anna, 99, 203
Kaiser, Florian, 69
Kenny, Meryl, 16
Keohane, Robert, 15
Ki-moon, Ban, 136
Knowledge, 98–99, 103, 104–106, 110–123, 126–129, 180–181, 199–200
and expertise called into question, 117–123
Knowledge broker, 28, 115–116, 125, 127
Kronsell, Annica, 9, 99, 203

Levin, Kelly, 3
Liverman, Diana, 130
Love Canal, 94
Lugones, María, 187–188

Maathai, Wangari, 142
MacGregor, Sherilyn, 9, 20, 55, 70, 112, 131, 135, 166, 179, 181, 203
Mackay, Fiona, 16
Magnusdottir, Gunnhildur Lily, 9
Malawi, 53
Maniates, Michael, 135
Marginalization, 6–7, 8, 21, 30, 31, 33, 39, 44–55, 58, 60–62, 131, 172–173, 176, 185–186, 188, 189, 193–196, 202
  based on age, 51
  based on class, 47–49
  based on disability, 49–51
  based on place, 52–54
  based on race, 44–47
Margolin, Jamie, 170
Marsh, David, 18
Masculinity, 5, 6, 8, 9, 10, 25, 40, 41, 42, 43, 82, 92, 100, 111, 112, 113, 126, 127, 129, 131, 159, 164, 168, 170, 179, 181, 183, 184, 186, 205
May, Theresa, 142
McCright, Aaron M., 34, 100
McKenna, Catherine, 170
Meijers, Marijn H. C., 182
Mellor, Mary, 67
Merchant, Carolyn, 67
Merkel, Angela, 183
Mexico, 33
Micheletti, Michele, 136
Mohanty, Chandra Talpade, 57
Moi, Toril, 32
Moosa, Christina Shaheen, 127
Morocco, 25
Motherhood narratives, 69, 70, 86–89, 93–94, 95, 96, 195–196
Mothers Out Front, 86

Natural disaster, 9, 10, 29, 30, 37–39, 46, 48–49, 50–51, 53, 54, 56, 59, 60, 62–63, 67, 78, 80, 82
Nazar, Nadia, 170
Nelson, Julie, 128
Nepal, 26, 105, 180, 183
Neumayer, Eric, 37
Neville, Kate, 18
New Zealand, 100
Nicaragua, 26
Niger, 149
Nigeria, 26
Nongovernmental organizations (NGOs), 1, 4, 30, 42–43, 63, 109, 183
  environmental organizations, 48, 86–88, 97, 137, 139, 148, 156, 161–164, 165, 169, 172–173, 188, 201–202, 205
Norway, 100

O'Brien, Karen, 135
Ocasio-Cortez, Alexandria, 170
O'Neill, Kate, 17
Otto, Siegmar, 69

Paris Agreement, 59, 69, 107, 108, 158
Participation, 9, 87, 106, 108–109, 111, 137–142, 148–155, 201–202
  women underrepresented in climate work, 43, 108–109, 148–151, 180, 185, 188
Partzsch, Lena, 136
Peru, 33
Pettenger, Mary E., 20, 176
Pew Research Center, 27
Philip, Kavita, 199
Plümper, Thomas, 37
Plumwood, Val, 182
Podoshen, Jeffrey, 146
Population growth, 83–86, 95, 179, 202
Poverty, 31, 46, 47–49, 51, 52–53, 57, 60, 62, 77, 84
Princen, Thomas, 143, 191

Pulé, Paul, 100
Puri, Lakshmi, 24

Race, 5, 6, 7, 21, 28, 32, 34, 42, 43, 44–47, 54, 58, 87, 100, 102, 121, 123, 130–131, 135, 148–149, 150, 153–154, 156, 157, 160, 161, 162–164, 167, 168, 172–173, 181, 188, 190, 193, 194, 195–196, 198, 201–202, 203
Reprocentricity, 70, 74, 94
Resource use, 36, 52, 54, 57, 63, 84, 106, 125, 172, 187, 200
Resurrección, Bernadette P., 32
Risk perception, 33–34, 100, 168, 178
Robinson, Fiona, 67
Robinson, Mary, 86, 142
Rural, 5, 30, 33, 48, 50, 52, 53, 54, 57, 105, 106, 125, 187, 189, 194
Rutjens, Bastiaan T., 182

Sandilands, Catriona, 69
Sasser, Jade, 84
Science, technology, engineering, and math (STEM), 119–123, 148, 158, 164, 165. *See also* Climate science
Scientific Revolution, 41
Scotland, 25, 51
Segal, Brenda, 146
South Korea, 27
Spain, 27
Stern, Maria, 20
Stolle, Dietlind, 136
Sultana, Farhana, 8, 82
Sweden, 100, 170

Taiwan, 25
Technology, 41, 110, 113, 119, 127, 130, 138, 141, 156–160, 162, 163, 164, 179, 181–183, 184
Tew, Madelaine, 170
Thøgersen, John, 69
Thomas, Kimberley, 63
Thunberg, Greta, 170, 171, 172
Tickner, J. Ann, 15
Tonkiss, Fran, 20
Tran, Dalena, 171
Tuana, Nancy, 127
Turner, Bryan, 32

UK Tar Sands Network, 172
United Church of Christ, 47
United Kingdom (UK), 27, 48, 49–50, 51, 54, 58, 63, 76–77, 97, 100, 104, 117, 151–152, 161, 172, 177
United Nations Conference on Environment and Development (Rio Earth Summit), 90, 97
United Nations Development Programme, 68
United Nations Environment Assembly, 60
United Nations Framework Convention on Climate Change (UNFCCC), 107, 108, 109, 148
United States (US), 25, 27, 34, 38, 46, 47, 48, 54, 69, 87, 89, 95, 100, 101, 116, 133, 137, 140, 146, 150, 154, 173, 194
United States Environmental Protection Agency, 94, 142

Venezuela, 25
Vietnam, 32, 153
Vulnerability, 2, 5, 9, 20, 26–27, 29–34, 37–39, 48, 49, 50, 51, 53, 54, 55, 57, 59–64, 69, 70, 77, 78, 80–83, 91, 97, 99, 128, 168, 176, 192, 193–195

Warner, Benjamin, 63
Whyte, Kyle Powys, 67
Wibben, Annick, 20
Wilson, Joanna, 179
Wirth, Kelsey, 86
Women and Gender Constituency, 109
Women-as-agents discourse, 133–173, 183, 201–204

Women-as-caregivers discourse, 38, 97, 103, 122, 155, 167, 187, 195–198
Women-as-knowledgeable discourse, 165, 187, 198–200
Women-as-vulnerable discourse, 34–64, 80, 93, 97, 105, 187, 193–195
Women's Environment and Development Organization (WEDO), 164
World Allergy Organization, 79

Xiao, Chenyang, 34

Young, Iris Marion, 32
Youth, 51, 91, 129, 169–170, 172, 194, 204–205